Advances in Applied Neurological Sciences 3

Nausea and Vomiting: Mechanisms and Treatment

Edited by
C. J. Davis, G. V. Lake-Bakaar
and D. G. Grahame-Smith

With 39 Figures

Springer-Verlag
Berlin Heidelberg New York Tokyo

Surgeon Lieutenant Commander CHRISTOPHER J. DAVIS BSc (Hons),
MB, BS, AKC, Royal Navy, Corpus Christi College, Oxford,
Senior Specialist in Applied Physiology (Royal Navy)
and Research Training Fellow, University Department of Clinical
Pharmacology, Radcliffe Infirmary, Oxford OX2 6HE,
United Kingdom

Dr. GERRY LAKE-BAKAAR BSc (Hons), MD, MRCP,
Research Gastroenterology Physician and Director of Medical Affairs,
Medical Department, Janssen Pharmaceutical Ltd (UK),
Wantage Oxfordshire, OX12 ODQ,
United Kingdom

Professor DAVID G. GRAHAME-SMITH MA, MB, BSc, PhD, FRCP,
Rhodes Professor of Clinical Pharmacology, University of Oxford,
Fellow of Corpus Christi College, Oxford,
Honorary Director of the MRC Clinical Pharmacology Unit,
Radcliffe Infirmary, Oxford and Honorary Consultant
Physician in General Internal Medicine and Clinical Pharmacology,
Oxford Health Authority, Oxford OX2 6HE,
United Kingdom

The meeting was sponsored by the Medical Department of Janssen Pharmaceutical Ltd (UK) and Janssen Pharmaceutica nv., Belgium

ISBN-13: 978-3-642-70481-9 e-ISBN-13: 978-3-642-70479-6
DOI: 10.1007/978-3-642-70479-6

Library of Congress Cataloging-in-Publication Data. Main entry under title: Nausea and vomiting. (Advances in applied neurological sciences; 3) Papers presented at an international symposium held in Oxford in 1984. Includes index. 1. Nausea – Congresses. 2. Vomiting – Congresses. I. Davis, C. J. (Christopher John), 1950– . II. Lake-Bakaar, G. V. (Gerry V.), 1948– . III. Grahame-Smith, David Grahame. IV. Series. [DNLM: 1. Nausea – congresses. 2. Vomiting – congresses. W1 AD436AH v. 3/WI 146 N313 1984] RB150.N38N38 1986 616'.047 86-902
ISBN-13: 978-3-642-70481-9(U.S.)

Softcover reprint of the hardcover 1st edition 1986

2125/3130-543210

Preface

The following papers were presented at an international symposium on the mechanisms and treatment of nausea and vomiting in man held in Oxford in 1984. I believe that this meeting was the first occasion on which representatives from such a wide variety of scientific and clinical specialities had come together to review and debate the spectrum of the vomiting phenomenon. An attempt was made to put before an invited international audience all the pertinent facts on the different facets of the topic and then to encourage extensive discussion of the contentious issues.

The first day of the meeting was devoted to the basic science approach to the problem and the second day to the more clinical aspects. This format has been broadly retained in the layout of the book, with the addition of summary chapters reviewing each day's contributions and focusing upon areas of particular importance. Acknowledgement must here be made to the enormous input from the many participants who either spoke in the debate or rose and themselves gave small presentations in addition to those of the invited speakers. The meeting occurred at a time of increasing interest in the problem of nausea and vomiting, especially because of its importance in cancer chemotherapy and radiation therapy, and even in space travel. That there is currently no consistently effective, reliable, acceptable and rational pharmacological solution to these problems lends weight to the argument that even after the 34 years which have elapsed since Wang and Borison first made the distinction between chemoreceptor trigger zone and vomiting centre, our level of understanding of the central control mechanisms for vomiting is still rudimentary.

It is my belief that this book, representing as it does a truly multidisciplinary approach to this vexatious problem, comes at the beginning of one of the most fruitful periods of research into the topic since the 1950s. Let us hope that it will give added impetus to this resurgence of interest and provide a useful reference work for those already involved with the problem both as clinicians and scientists.

We are grateful to Heather Waldron MSc., Judith Alexander and Anne Brooks of the Medical Department, Janssen Pharmaceutical Limited (UK), for their excellent efforts towards the organisation and co-ordination of the symposium, and the preparation of this manuscript.

February, 1986

C. J. Davis
Corpus Christi College, Oxford

Contents

List of Contributors

You will find the addresses at the beginning of the respective contribution

The Multiple Causes of Vomiting: Is There a Common Mechanism?

D. G. Grahame-Smith [1]

It is common clinical experience that many different pathologies may cause vomiting. The variety of these is illustrated thus:

Gastro-intestinal diseases: gastritis, gastric irritants, intestinal obstructions, intra-abdominal inflammation, hepatitis, pancreatitis, enteritis
Infections ("toxic" state): viral, bacterial
Pain
Metabolic diseases: uraemia, Addison's disease, hypercalcaemia, diabetic ketoacidosis
Intra-cranial disease: increased intra-cranial pressure (tumours, intra-cranial haemorrhage), malignant hypertension
Labyrinthine and vestibular disorders: motion sickness, labyrinthitis, vestibular neuronitis
Pregnancy: early morning sickness, hyperemesis gravidarum
Exogenous emetic substances: poisons and drugs
Radiation sickness
Psychological: terror, fright, disgust "conditioned reflex"
Migraine
Circulatory syncope

All of these different causes may result in the complex sequence of physiological events which precede or accompany the act of vomiting. These include nausea, salivation, yawning, and the co-ordinated respiratory, gastro-intestinal and abdominal muscular movements which result in retching and vomiting, accompanied by changes in body posture, cardiovascular function and psychological state.

One has only to consider the wide spectrum of causes to guess intuitively that there cannot be just one mechanism through which they all originate.

Most authorities now divide the emetic process into three components:

1. The afferent pathways to the central nervous system (CNS) relaying the signals of emetic stimuli
2. The reception, recognition and central processing of these signals leading eventually to integrated emetic efferent signals emerging from the CNS
3. The efferent pathways relaying the signals which lead to the co-ordinated respiratory, gastro-intestinal and abdominal muscle expulsive movements and the accompanying emetic epiphenomena

1 MRC Clinical Pharmacology Unit, University Department of Clinical Pharmacology, Radcliffe Infirmary, Oxford OX2 6HE, UK

Nausea and Vomiting: Mechanisms and Treatment
Ed. by C. J. Davis, G. V. Lake-Bakaar and D. G. Grahame-Smith

Afferent Pathways

From the Gastro-Intestinal Tract

Vomiting and retching can be elicited in experimental animals by stimulation of the cut central ends of the abdominal vagal branches to the stomach (Andrews et al. 1985; see also Borison and Wang 1953). Vomiting produced by the distension of pyloric pouches can be abolished by transthoracic vagotomy. The local "irritative" emetic action of nitrogen mustard on the stomach can be prevented by vagotomy, but not by cutting the splanchnic nerves. The vomiting due to copper sulphate requires intact vagal and splanchnic afferents, the former being more important. It appears that the afferents involved in vomiting accompanying intestinal obstruction are sympathetic (Sharma et al. 1972).

Vagal and Sympathetic Afferents Arising from Other Viscera

Biliary tract and gall bladder disease, peritoneal inflammation and cardiac pain may cause vomiting and the vagal and sympathetic afferents from these organs relay emetic signals, although vomiting associated with cardiac causes appears to be mediated mainly by vagal afferents (Abrahamsson and Thoren 1973).

These few examples clearly demonstrate that either vagal or sympathetic afferents carry emetic stimuli of visceral origin.

Chemoreceptor Trigger Zone Input

In a series of classical experiments, Borison and Wang (see review 1953) discovered an area on the floor of the fourth ventricle, within the area postrema (AP) which, if ablated, rendered dogs refractory to the vomiting stimulus of apomorphine, whilst retaining responsiveness to intra-gastric copper sulphate. The AP presents an incomplete blood-brain barrier. The idea was proposed that there is a superficial area of the AP bathed in CSF, relatively unprotected by the blood-brain barrier, which receives chemical signals and which passes them on to other deeper structures in the medullary reticular system through polysynaptic neuronal links. Borison and Brizzee (1951) originally defined the chemoreceptor trigger zone (CTZ) in the cat as a bilateral portion of tissue less than one cubic millimetre in size on either side of the fourth ventricle, contiguous medially with the area postrema.

Not a great deal appears to be known about how the chemoreceptor functions of the CTZ actually operate. What are the chemosensitive cells? Why are they chemosensitive and what are the chemosensitive receptors and how do they work? Excatly how do the emetic stimuli reach them?

There is little doubt that the AP itself is important in man. Lindstrom and Brizzee (1962) lesioned the AP of five patients with intractable vomiting due to inoperable brain tumours who were then relieved of their vomiting and who became refractory to apomorphine. That study does not actually prove the importance or specificity of the chemoreceptor trigger zone in man but undoubtedly there is some function in the AP of man subserving vomiting; the surgical lesion may of course have covered both the CTZ and other underlying neural structures involved in vomiting.

I find the chemoreceptor function of the CTZ difficult to understand for the following reasons:

a) The molecules apparently acting through the CTZ to produce vomiting have diverse structures. For instance, there is apomorphine, morphine, ipecacuanha (composed of emetine and cephaeline alkaloids), cardiac glycosides, nicotine, L-dopa, cisplatinum, and a host of cytotoxic agents of different structures. One has to ask, "How can the CTZ sense such different structures? Is its function like that of the olfactory epithelium subserving the sense of smell?" (Not that such a consideration helps very much as the olfactory mechanisms themselves are also rather obscure.)
b) While I can understand intuitively that the CTZ senses toxins quickly, leading to their ejection from the stomach, and that therefore this is a good evolutionary reason for having a CTZ, why is it also involved in motion sickness and radiation sickness, assuming that ablation experiments have been specific and have not damaged afferent emetic relaying fibres? How or why then does the CTZ have a dual role as a chemical trigger mechanism and as a facilitatory mechanism in some other causes of vomiting? The concept of the chemoreceptor trigger zone is of vital importance when considering the action of antiemetic drugs. What is the direct evidence that they affect the CTZ? It is true that the dopamine antagonists block the action of apomorphine, but are they antagonising apomorphine at the CTZ? Do they act at some other point in the neural processes mediating vomiting in which dopamine is involved?

It is presumed that from the CTZ neural connexions exist which carry the emetic stimuli to those areas of the lateral medulla termed "the vomiting centre" (see below).

From the Labyrinth

Motion sickness, labyrinthitis and that rather obscure syndrome of vestibular neuronitis are all associated with vomiting and its galaxy of epiphenomena. There seems little doubt that labyrinthine stimulation leads to impulses passing along the vestibular nerve to the lateral medulla, causing emesis (Reason and Brand 1975). There is also evidence that the vestibular cerebellum is involved in motion sickness as extirpation of the cerebellar nodulus prevents motion sickness in the dog (Wang 1965).

Afferents from Other Areas of the Central Nervous System

Electrical stimulation of certain areas of the hypothalamus in the cat can occasionally induce vomiting which may be delayed (Borison and Wang 1953). Other authors will write in the symposium proceedings of the importance of psychological factors in nausea and vomiting and one presumes, therefore, that higher cerebral function provides afferent signals to the vomiting centre.

The Integrated Vomiting Centre

The next component of the emetic process to consider is "the vomiting centre". The idea of a vomiting centre precedes the experiments of Borison (1948), Borison and Wang (1949) and Borison (1959), but they were the first workers to be able to induce vomiting in cats by stimulating certain areas of the lateral reticular formation of the medulla. Stimulation of other areas around what they termed the vomiting centre produced salivation, spasmodic respiratory movement, forced inspiration and vasomotor reactions. Close by are the vestibular nuclei and the reticular activating system. Ablation experiments showed that elimination of that area inhibited vomiting in response to various stimuli. Recently, Miller and Wilson (1983) have re-examined the vomiting centre concept using electrical stimulation of the brain stem of decerebrate cats. They were, however, unable to define unequivocally a localised and specific vomiting centre. Crucial to Borison and Wang's (1949) concept of a vomiting centre was that the centre itself does not carry out the function of vomiting, but it co-ordinates the activities of other neural structures in its immediate vicinity to produce a complicated patterned and stereotyped response.

Miller and Wilson (1983) believe that their results are more consistent with the concept that the neurones involved in the control of vomiting are diffusely distributed in the region described by Borison and Wang (1953). No one will deny that the lateral medulla is important in the production of vomiting, but whether there is a discrete centre from which the vomiting act is programmed is, on the basis of the findings of Miller and Wilson (1983), open to further investigation.

This is important in the consideration of precisely where anti-emetic drugs might act or might be developed to act. Basically, although it is believed that a neural network in the lateral medulla is activated by afferent emetic stimuli, nevertheless complex respiratory and somatic muscle movement, together with autonomic epiphenomena and gastro-intestinal movements, have still then to be integrated to produce emesis.[1]

Efferent Systems

The final component mediating the activity of the lateral medulla in initiating the vomiting act consists of somatic efferent nervous pathways controlling respiratory (Monges et al. 1974) and abdominal musculature and visceral efferent components mediating changes in gastric tone and motility (Abrahamsson 1973; Ab-

1 Throughout this volume, the terms vomiting and emesis are used interchangeably

rahamsson et al. 1973) and the autonomic epiphenomena such as salivation, pallor, and sweating. In this the vagus plays a prominent role, although exactly what the neurotransmitters are at the gastro-intestinal musculature is uncertain. Various substances, such as vasoactive intestinal polypeptide (VIP), purine nucleotides and dopamine have been suggested. It is important, though, to note the involvement of both autonomic and somatic pathways. These considerations show the potential complexity of the neural mechanisms underlying vomiting.

It is likely that a large number of neurotransmitter agents are involved within the overall neural connections. A great deal of work has to be done before coordinated neuropharmacological sense can be made of it and before we can point to the sites of action of anti-emetic and emetic drugs.

Emetics, Anti-Emetics and the Vomiting Control Mechanism

Stimuli which cause emesis include cancer chemotherapeutic agents, apomorphine, morphine, digitalis, veratrum alkaloids, emetine, radiation, oral copper sulphate, and motion sickness.

Consideration of the variety of emetic stimuli listed above is sufficient illustration of the complexity of the stimulus-response relationships involved (Wang 1965).

Dopamine blocking agents, such as the phenothiazines, or butyrophenones, and drugs such as metoclopramide and domperidone, are very effective in blocking apomorphine-induced vomiting, but relatively ineffective against digitalis-induced vomiting. It does not seem that it is possible in man to clearly categorise emetic stimuli neuropharmacologically on the basis of the anti-emetic drugs which are effective against them.

Radiation-induced sickness poses interesting problems since the chemoreceptor trigger zone seems to be important for its early phase, perhaps being stimulated by substances released from irradiated cells (Wang 1965). In the later phases of radiation-induced sickness, gastro-intestinal inflammatory change with stimulation of abdominal visceral afferents may be important (Wang 1965). The early and later vomiting induced by nitrogen mustard might have similar mechanisms. In the uraemic syndrome in dogs, chronic CTZ ablation attenuated vomiting and a plausible hypothesis is that some toxin accumulating in renal failure might be responsible for the emesis, but the picture is not yet clear (Wang 1965).

Emetine-induced vomiting is particularly interesting in relation to the possible role of protein synthesis in nausea and vomiting. Emetine inhibits brain protein synthesis and brain protein synthesis is involved in neurotransmitter function (Grahame-Smith 1972; Green et al. 1976; Grahame-Smith and O'Shaughnessy 1985). Could it be that inhibition of protein synthesis in the AP or a closely related medullary area is emetic? Harris (1982) proposed that chemotherapy-induced vomiting might inhibit enkephalin degrading enzymes in the CTZ, where it was proposed enkephalin excited emetic receptors, perhaps through some dopamine neuronal interaction. If enkephalin degrading enzymes were inhibited, then enkephalin levels would rise and vomiting would be induced. There is as yet no proof of this hypothesis, but it is now readily open to investigation.

The problem of specific neurotransmitter mediation of chemotherapy-induced emesis or any of the other causes of emesis, is complex and difficult to interpret. Peroutka and Snyder (1982) pointed out that anti-emetic drugs variously exhibited histamine (H_1), muscarinic (cholinergic) and/or dopamine (D_2) blocking actions, and proposed that the concurrent blockade of these three receptor types would provide the best anti-emetic activity, presumably by interfering with those particular neurotransmitter functions in central emetic mechanisms. However, when, for instance, a dopamine blocking agent blocks the emetic effect of apomorphine, is it acting at some apomorphine/dopamine "sensory" receptor in the CTZ or blocking some "downstream" pathway. Moreover, if the latter is the case, why are the dopamine blocking agents not generally more effective against other emetic stimuli?

The anti-emetic drugs can be classified as:

1. Anti-dopamine agents, e.g. phenothiazines, butyrophenones, metoclopramide and domperidone
2. Histamine (H_1) blocking agents
3. Muscarinic anti-cholinergic agents, e.g. hyoscine
4. Cannabinoids and related agents
5. Cortico-steroids (in chemotherapy-induced vomiting)

As mentioned above, the actual sites in the emetic pathways where dopamine, histamine (H_1) and cholinergic muscarinic blockade are active are not known.

The fact that domperidone does not easily cross the blood-brain barrier probably puts its central anti-dopamine action somewhere around the AP, but one cannot be sure. There is no satisfactory explanation of the anti-emetic action of cannabinoids.

With the advent of metoclopramide and domperidone, and assuming that part of the anti-emetic effect of agents is mediated peripherally through their anti-dopamine action, interest has been shown in dopaminergic mechanisms hastening stomach emptying, increasing oesophageal sphincter activity and increasing the movement of gut contents along the upper gastro-intestinal tract (Valenzuela 1976; Lanfranchi et al. 1978; Ennis et al. 1978; Cools 1978). It is supposed that these actions are useful in overall anti-emesis activity. Both metoclopramide and domperidone may act on the chemoreceptor trigger zone and their actions there and peripherally are difficult to dissect. It is not clear to me how important the peripheral anti-dopamine effects of domperidone and metoclopramide are.

Conclusion

Consideration of the neural afferent and efferent pathways, the CTZ and medullary structures responsible for the co-ordinated emetic act coupled with the large number of different emetic stimuli, which do not have a commonality, shows clearly that any simple idea of a unitary mechanism underlying vomiting is invalid. It is not surprising, therefore, that no single anti-emetic drug is efficacious in all vomiting syndromes and unless there is some crucial specific neurotransmitter subserving a final common pathway within the medullary structures responsible for vomiting, an all-embracing and effective anti-emetic does not seem a likely possibility. However, if some specific neurotransmitter with its associated receptors was found which was vitally important for integrating all the different signals involved in vomiting and processing them into vomiting efferent signals, then that possibility would be a viable one.

References

Abrahamsson H (1973) Studies on the inhibitory nervous control of gastric motility. Acta Physiol Scand [Suppl] 390

Abrahamsson H, Thoren P (1973) Vomiting and reflex vagal relaxation of the stomach elicited from heart receptors in the cat. Acta Physiol Scand 83:433–439

Abrahamsson H, Jansson G, Martinson J (1973) Vagal relaxation of the stomach induced by apomorphine in the cat. Acta Physiol Scand 88:296–302

Andrews PLR, Bingham S, Davis CJ (1985) Retching evoked by stimulation of abdominal vagal afferents in the anaesthetised ferret. J Physiol (Lond) 358:103

Borison HL (1948) Electrical stimulation of the central nervous mechanism for vomiting in the cat. Am J Physiol Proc 155:428

Borison HL (1959) Effect of ablation of medullary emetic chemoreceptor trigger zone on vomiting responses to cerebral intraventricular injection of adrenalin, apomorphine and pilocarpine in the cat. J Physiol (Lond) 147:172–177

Borison HL, Brizzee KR (1951) Morphology of emetic chemoreceptor trigger zone in cat medulla oblongata. Proc Soc Exp Biol Med 77:38–42

Borison HL, Wang SC (1949) Functional localization of central co-ordinating mechanism for emesis in the cat. J Neurophysiol 12:305–313

Borison HL, Wang SC (1953) Physiology and pharmacology of vomiting. Pharmacol Rev 5:193–230

Cools AR (1978) Haloperidol and the significance of α-NE mediated control of a subpopulation of DA receptors for concepts such as supersensitivity and tolerance: a behavioral study in cats. Life Sci 23:2475–2483

Ennis C, Schneiden H, Cox B (1978) Effects of the antiemetic drug domperidone on guinea pig isolated gastrointestinal tissue. J Pharm Pharmacol 30 [Suppl]:34

Grahame-Smith DG (1972) The prevention by inhibitors of brain protein synthesis of the hyperactivity and hyperpyrexia produced in rats by monoamine oxidase inhibition and the administration of L-tryptophan or 5-methoxy-N,N-dimethyl tryptamine. J Neurochem 19:2409

Grahame-Smith DG, O'Shaughnessy KM (1985) Evidence that inhibition of brain protein synthesis inhibits striatal dopamine release. Br J Pharmacol 84:50 P

Green AR, Heal DJ, Grahame-Smith DG, Kelly PH (1976) The contrasting actions of TRH and cycloheximide in altering the effects of centrally acting drugs: evidence for the non-involvement of dopamine sensitive adenylate cyclase. Neuropharmacology 15:591–599

Harris AL (1982) Cytotoxic-therapy-induced vomiting is mediated via enkephalin pathways. Lancet 1:714–716

Lanfranchi GA, Marzio L, Cortini C, Trento L, Labo G (1978) Effect of dopamine on gastric motility in man: evidence for specific receptors. In: Duthie HL (ed) Proceedings of the 6th international symposium on gastrointestinal motility. Edinburgh, 1977. MTP Press, Lancaster, pp 161–171

Lindstrom PA, Brizzee KR (1962) Relief of intractable vomiting from surgical lesions in the area postrema. J Neurosurg 19:288

Miller AD, Wilson VJ (1983) Vomiting centre reanalyzed: an electrical stimulation study. Brain Res 270:154–158
Monges H, Salducci J, Naudy B (1974) Electrical activity of the gastrointestinal tract in the dog during vomiting. In: Daniel EE (ed) Gastrointestinal motility. Mitchell, Vancouver, pp 479–488
Peroutka SJ, Snyder SH (1982) Antiemetics: neurotransmitter receptor binding predicts therapeutic actions. Lancet 1:658–659
Reason JT, Brand JJ (1975) (eds) Motion sickness. Academic, New York
Sharma RN, Dubey PC, Dixit KS, Bhargava KP (1972) Neural pathways of emesis associated with experimental intestinal obstruction in dogs. Indian J Med Res 60:291–295
Valenzuela JE (1976) Dopamine as a possible neurotransmitter in gastric relaxation. Gastroenterology 71:1019
Wang SC (1965) Emetic and antiemetic drugs. In: Root WS, Hofmann FG (eds) Physiological pharmacology, vol 2. Academic, New York, pp 255–328

Section I.
Anatomy, Physiology, and Pharmacology

Anatomy and Physiology of the Chemoreceptor Trigger Zone and Area Postrema

H. L. BORISON [1]

Identification of the Area Postrema as the Emetic Chemoreceptor Trigger Zone

Finding the Vomiting Centre

It was not until the vomiting centre itself was correctly localized in the bulbar reticular formation that the idea of a separate chemosensory station could be formulated (Borison and Wang 1953). Figure 1 shows where the vomiting centre was thought to be located before 1949 (panel A) and then where it was mapped by Borison and Wang (1949) (panel B). Earlier workers (Hatcher and Weiss 1923)

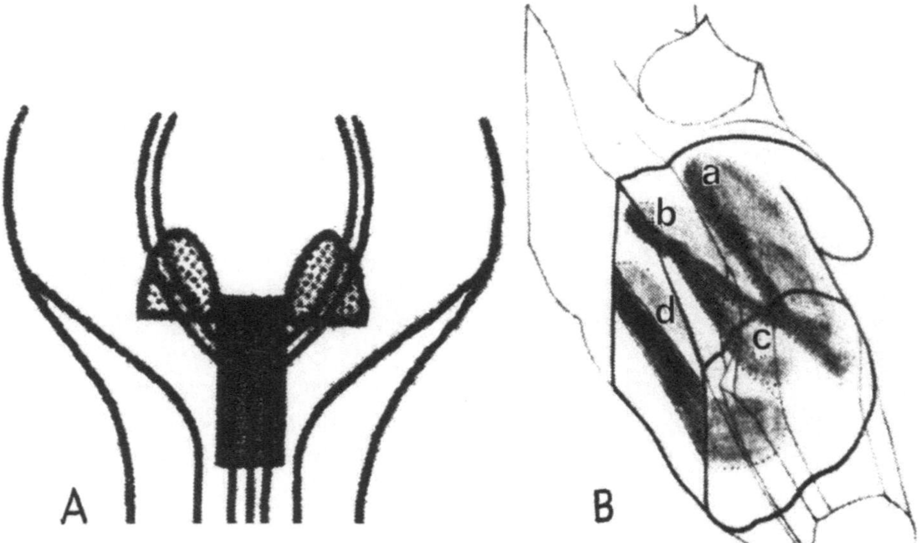

Fig. 1. *Panel A.* Drawing copied from Hatcher and Weiss (1923) showing their location of the vomiting centre bilaterally in the region of the vagal sensory nucleus, compared with the prior mid-line location (*rectangle*) designated by Thumas. *Panel B.* Phantom view of electrically responsive regions in the cat hemi-medulla showing the vomiting centre (*c*), surrounded by centres for spasmodic respiratory movement (*a*), salivation (*b*) and forced inspiration (*d*) (reproduced from Borison and Wang 1949)

1 Department of Pharmacology and Toxicology, Institute of Brain Stem Studies, Dartmouth Medical School, Hanover, New Hampshire, 03755, USA

Nausea and Vomiting: Mechanisms and Treatment
Ed. by C. J. Davis, G. V. Lake-Bakaar and D. G. Grahame-Smith

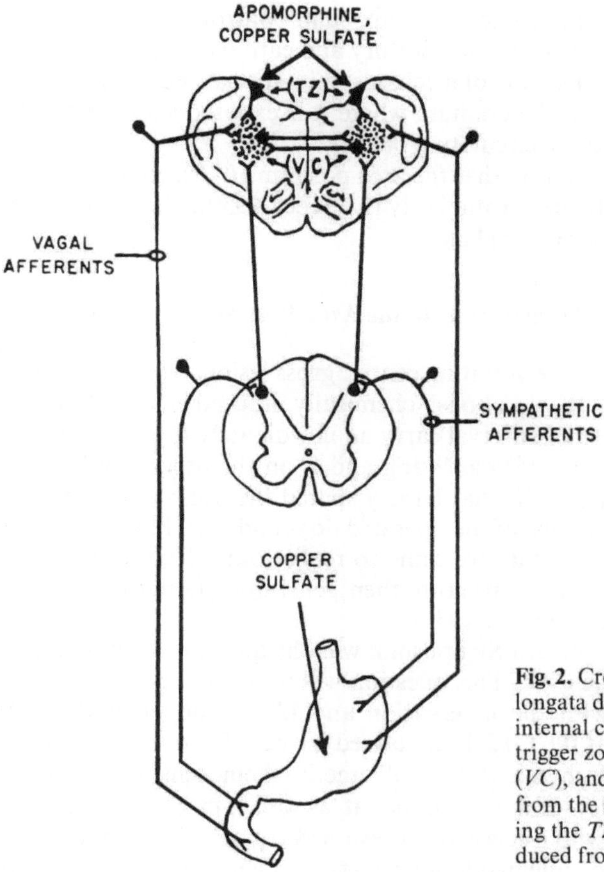

Fig. 2. Cross-section of the medulla oblongata diagrammatically showing the internal connection of the chemoreceptor trigger zone (*TZ*) to the vomiting centre (*VC*), and the visceral afferent nerves from the stomach and duodenum bypassing the *TZ* on their way to the *VC* (reproduced from Wang and Borison 1952)

simply assumed that the vomiting centre was inherently responsive to "centrally-acting" emetic drugs applied locally to the ala cinerea on the brain surface; whereas, we used direct electrical stimulation to activate the centre in the parenchymal reticular formation. Indeed, there is at present no neuropharmacological basis for attributing any functional chemoselectivity to the particular medullary neurons that comprise the vomiting centre.

Formulating the Chemoreceptor Trigger-Zone Concept

Our formulation of the chemoreceptor trigger-zone (CTZ) concept was based on three main lines of evidence, namely: (a) the classic emetic chemosensitivity of a structure on the dorsal medullary surface; (b) the deep location of the vomiting centre; and (c) the retention of emetic capability in animals subjected to surgical ablation of the chemosensitive locus. This concept is illustrated in Fig. 2, as it was originally depicted, showing afferent nerve fibres from the gut entering the vomiting centre through pathways that bypass the trigger zone. Thus, the CTZ is viewed as one among many inputs to the vomiting centre.

Peripheral chemosensory inputs to the brain-stem control centres were, of course, already well established for the respiratory and cardiovascular reflex systems. In the case of the vomiting control mechanism, its specialized chemosensor (i.e. the CTZ) is located within the cranium where it is exposed simultaneously to the cerebrospinal fluid and the circulating blood.

There has even been a suggestion that the area postrema (AP), containing the CTZ, resembles histologically the carotid body (De Kock 1959). Curiously, both organs were at one time described as glands.

Assigning the Chemoreceptor Trigger Zone to the Area Postrema

In the early attempts to locate the vomiting centre, gross lesions of the vagal nuclear complex were made acutely to abolish chemically induced emesis. Animals thus prepared for the purpose of survival carry an inordinately high risk of premature death from pulmonary oedema. Wang and I, on the other hand, made minimal lesions to destroy the CTZ that largely spared the dorsal vagal nuclei, thereby assuring indefinite survival of the operated dogs and cats. It was not until I joined forces with KR Brizzee that we came to realise our effective lesions of the CTZ intruded upon the AP, a structure then generally ignored by medical neuroanatomists (Borison and Brizzee 1951).

Of special interest to the present Symposium was the question whether all or part of the AP constitutes the CTZ. This question will be given further consideration later in connection with the innervation and blood supply of the AP. Nonetheless, it is agreed that the CTZ is contained in the AP alone and that the underlying medullary parenchyma is free of emetic chemosensory elements. Figure 3 presents a dorsal view of the AP in the cat, as well as cross-sections from a normal animal compared with a successfully lesioned animal. To the best of my knowledge, the only such documented lesions made in the human were reported by Lindstrom and Brizzee (1962). The topography of the calamus scriptorius in man bears a striking similarity to that in the dog, cat and monkey.

The Vomiting Control Mechanism

The act of vomiting is programmed in the lower brain stem. The midbrain decerebrate animal is capable of performing the entire program of emetic behaviour which surprisingly includes the premonitory signs of nausea. The full behavioural syndrome of motion sickness has been observed in a decorticate human being (Doig et al. 1953).

It would appear, therefore, that the mental distress of nausea associated with vomiting results secondarily to the fundamental subconsciously controlled emetic process. It also follows that while relief from the suffering of nausea might be afforded with mind-altering drugs, general non-specific therapeutic blockade of the vomiting act requires direct intervention in the brain-stem control process. Specific vomiting responses initiated from identified emetic sensory receptors are theoretically vulnerable to interruption at their afferent points of origin.

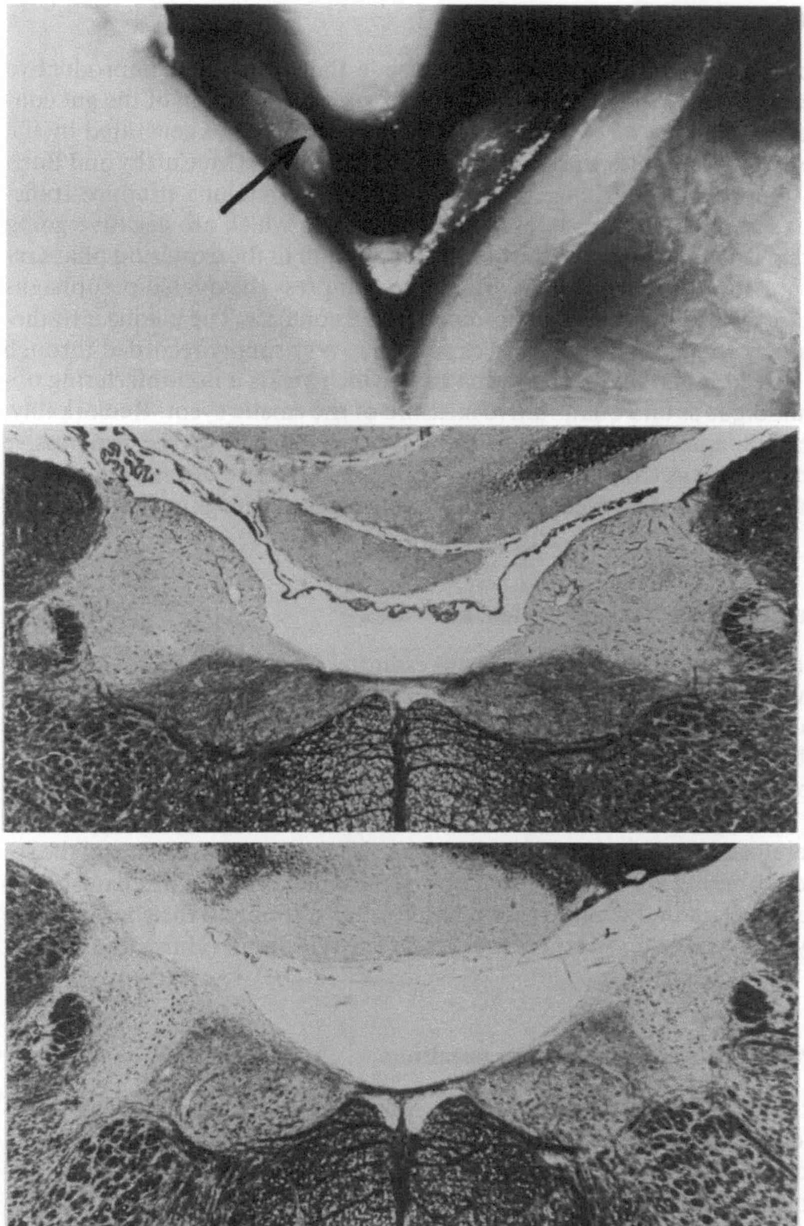

Fig. 3. *Top panel.* Calamus scriptorius of cat medulla oblongata with area postrema (*arrow*) on lateral margins bulging into the fourth ventricle. *Middle* and *bottom panels,* respectively: Cut through mid-section of intact area postrema compared with chronic lesion of same. Stained with luxol fast blue, cresyl violet and Periodic Acid-Schiff technique

The Somatic Mechanics of Vomiting

The vomiting act is typically a two-step episode that consists of unproductive retching terminated by the forceful expulsion of vomitus. Ejection of the gut contents is the consequence of large pressures in the body cavities generated by the somatic respiratory muscles working against a closed glottis (McCarthy and Borison 1974). Opposed synchronized contractions during retching produce transdiaphragmatic pressure pulses in excess of 200 mmHg which are negative-going in the thorax. A sudden upward shift of the diaphragm in the expulsion phase reverses the pressure within the thorax finally to compress the dilated oesophagus with resultant evacuation of the gastro-oesophageal contents. The unique intrathoracic pressure pattern of retching and expulsion is very simply recorded through a catheter placed in the thin-walled vena cava which yields a non-interfering objective, unambiguous and quantitative measure of the emetic event. Remarkably, repeated emetic episodes retain their stereotyped form despite the progressive evacuation of vomitus from the stomach. The somatic nature of the vomiting act is underlined by the ineffectiveness of autonomic blocking drugs in altering its expression.

Role of the Autonomic Nervous System in Nausea

While the autonomic nervous system is not essential for the mechanical act of vomiting, involvement of those visceral efferent nerves in the total emetic process appears to be directly proportional to the duration and intensity of accompanying nausea. Thus, the occurrence of sweating, pallor, salivation, gastric stasis, diarrhoea, hypotension, tachycardia, etc. results from altered activities of the sympathetic and parasympathetic divisions of the autonomic nervous system.

It is therefore apparent that nausea, with its visceral concomitants, and the somatic act of vomiting are distinct functions of the emetic control mechanism. Indeed, rodent species that cannot vomit manifest most if not all the characteristic psychophysiological and visceral responses to emetic stimuli (Hulse and Patrick 1977).

Emetic Drugs, Receptors and Neurotransmitters

A number of neurotransmitters, i.e. amines, amino acids and peptides, cause vomiting when administered as drugs (Carpenter et al. 1984). Since it has not been possible to interrupt central impulse transmission in the emetic reflex arc selectively with any known postsynaptic receptor-blocking agent, it is difficult to imagine that emesis-inducing neurotransmitters could activate the vomiting control mechanism in mid-stream. More likely, those agents act in the manner of other emetic substances simply to excite matching sensory chemoreceptors whose specific chemical blockade does not interfere with the initiation of vomiting through parallel unblocked receptor inputs. Furthermore, the excitation of sensory chemoreceptors in the area postrema by known neurotransmitters sheds no light on the nature of the actual transmitters operating in the synaptic connections of the emetic controller.

Relationship of the Area Postrema to the Blood-Brain Barrier

The Circumventricular Organ System

The AP is counted among a mixed group of spongiform vascular bodies, distributed along the ependymal-pial seam on the brain surface, which become stained with systemically injected dyes that pass through their leaky capillaries. This permeability characteristic is shared with the choroid plexus. Unlike the choroid plexus, however, the AP directly exchanges solutes with the cerebrospinal fluid by means of free flow through pericapillary spaces which are continuous with the subarachnoid fluid space. Because in our view the AP is not strictly part of the brain substance to which it is attached, it is more correct to say this vascular appendage lacks a blood-cerebrospinal fluid barrier rather than a blood-brain barrier.

The heterogeneous group of hypependymal vascularized structures has come to be known collectively as the circumventricular organ system. A collective function has not, however, been established for the "system" although a general neuroendocrine role has been suggested (Weindl 1973). The AP would serve in such a role as a receptive site of humoral communication through the cerebrospinal fluid.

Blood Supply and Innervation

The configuration of its blood supply and innervation are crucial elements in the surgical anatomy of the AP (Borison 1974). The organ has an essentially independent blood supply arising as a separate branch of the posterior inferior cerebellar artery. The vascular supply and associated nerves enter the AP from the pia mater on the subarachnoid side of the tela choroidea along the same line used as the guide for applying the surgical cautery to destroy the CTZ. A prominent nerve bundle constituting in all probability the major link to the emetic controller is immediately subjacent to the pial margin. Thus, essential routes of sustenance and communication for the entire AP are critically exposed to surgical interruption at the lateral border of that organ. Furthermore, a spinal pathway of visceral afferent nerves to the lower brain stem enters superficially at the nearby junction of the nucleus gracilis and the nucleus of the solitary tract (Kuru 1956).

Where Are the Chemoreceptors?

The architectural design of the AP may be viewed as a three-compartmental complex wherein the core blood compartment exchanges solute through fenestrations into its surrounding subarachnoid fluid compartment formed by an investiture derived from the pia, and the latter is separated from the ventricular fluid compartment by a selectively permeable membrane formed by the ependyma. Much uncertainty lies in the manner and extent to which ventricular fluid comes in contact with postremal parenchyma, that is, with the neural and glial elements of the organ, and in the dynamics of water and solute exchange between compartments. We know that emesis is evoked with a shorter latency by chemical delivery through the blood stream than by direct chemical application to the brain surface. We know that certain agents can evoke emesis only when injected into the blood compartment, e.g. puffer fish poison. We do not know in which compartment the chemoreceptors are located or whether they exist as free nerve endings, glial end-feet or specialized receptor cells. Neural cells of one or two unique type(s) are contained in the AP. However, the exact cellular origins of an internal nerve fibre network and the external communicating nerve tract have not been established definitively. The functional significance of neurotransmitter vesicles in synaptic nerve endings is not understood. The barrier nature of the postremal-brain interface remains obscure.

Other Functions of the Area Postrema

The AP has been implicated in a variety of physiological and behavioural responses to appropriate stimuli that did not evoke vomiting per se in the particular circumstances of the given experiments. On the other hand, it should be appreciated that the broad spectrum of signs associated with nausea and the residual physiological concomitants of inhibited vomiting in the anaesthetized state are in fact emesis-related phenomena. The following are non-emetic functional influences attributed to the AP: cardiovascular effects of angiotensin, serotonin-induced electroencephalographic arousal, salt and water excretion, blood glucose regulation, defaecation, conditioned taste aversion, inhibition of rumination, and gastric secretion and motility (see reviews by Borison 1974, and Borison et al. 1984a, b).

Debated Emetic Roles of the Area Postrema

Two therapeutic concerns of intense current public interest are the amelioration of motion/space sickness and of radiation sickness. Rational anti-emetic therapy requires the understanding of pathophysiological mechanism in each disorder. If the AP is an essential element in those emetic syndromes, then a humoral factor may be presumed to mediate the particular reflex process for that case. If, to the contrary, the AP is not involved, therapeutic strategies aimed at blocking a humoral factor at the CTZ would be to no avail. As things stand, disparate results have been reported following ablation of the AP in studies on both motion-induced emesis (Wang and Chinn 1954; Brizzee et al. 1980; Borison et al. 1984a, b) and radiation-induced emesis (Chinn and Wang 1954; Brizzee et al. 1955; Borison 1957). Reconciliation of these differences will undoubtedly come from a neurosurgical resolution of the problem.

References

Borison HL (1957) Site of emetic action of x-radiation in the cat. J Comp Neurol 107:439–453

Borison HL (1974) Area postrema: chemoreceptor trigger zone for vomiting – is that all? Life Sci 14:1807–1817

Borison HL, Brizzee KR (1951) Morphology of emetic chemoreceptor trigger zone in the cat medulla oblongata. Proc Soc Exp Biol Med 77:38–42

Borison HL, Wang SC (1949) Functional localization of central co-ordinating mechanism for emesis in cat. J Neurophysiol 12:305–313

Borison HL, Wang SC (1953) Physiology and pharmacology of vomiting. Pharmacol Rev 5:193–230

Borison HL, McCarthy LE, Borison R, Mandal AK, Fisk TJ (1984a) Motion sickness is not prevented by chronic ablation of area postrema in cats. Fed Proc 43:504

Borison HL, Borison R, McCarthy LE (1984b) Role of the area postrema in vomiting and related functions. Fed Proc 43:2955–2958

Brizzee KR, Neal LM, Williams PM (1955) The chemoreceptor trigger zone for emesis in the monkey. Am J Physiol 108:659–662

Brizzee KR, Ordy JM, Mehler WR (1980) Effect of ablation of area postrema on frequency and latency of motion induced emesis in the squirrel monkey. Physiol Behav 24:849–853

Carpenter DO, Briggs DB, Stominger N (1984) Peptide-induced emesis in dogs. Behav Brain Res 11:277–281

Chinn HI, Wang SC (1954) Locus of emetic action following irradiation. Proc Soc Exp Biol Med 85:472–474

DeKock LL (1959) The carotid body system of the high vertebrates. Acta Anat 37:265–279

Doig RK, Wolf S, Wolff HG (1953) Study of gastric function in a "decorticate" man with gastric fistula. Gastroenterology 23:40–44

Hatcher RA, Weiss S (1923) Studies on vomiting. J Pharmacol Exp Ther 22:139–193

Hulse EV, Patrick G (1977) A model for treating post-irradiation nausea and vomiting in man: the action of insulin in abolishing radiation-induced delay in gastric emptying in the rat. Br J Radiol 50:645–651

Kuru M (1956) The spino-bulbar tracts and the pelvic sensory vagus. Further contributions to the theory of the sensory dual innervation of the viscera. J Comp Neurol 104:207–231

Lindstrom PA, Brizzee KR (1962) Relief of intractable vomiting from surgical lesions in area postrema. J Neurosurg 19:228–236

McCarthy LE, Borison HL (1974) Respiratory mechanics of vomiting in decerebrate cats. Am J Physiol 226:738–743

Wang SC, Borison HL (1952) A new concept of organization of the central emetic mechanism: recent studies on the sites of action of apomorphine, copper sulphate and cardiac glycosides. Gastroenterology 22:1–12

Wang SC, Chinn HI (1954) Experimental motion sickness in dogs. Functional importance of the chemoceptive emetic trigger zone. Am J Physiol 178:111–116

Weindl A (1973) Neuroendocrine aspects of circumventricular organs. In: Ganong WF, Martini L (eds) Frontiers in neuroendocrinology. Oxford University Press, New York, pp 3–31

Neurochemical Investigations into the Human Area Postrema

J-C. Schwartz[1], Y. Agid[2], M-L. Bouthenet[3], F. Javoy-Agid[2], C. Llorens-Cortes[1], M-P. Martres[1], H. Pollard[3], N. Sales[3], and H. Taquet[2]

Introduction

It is now well established that the area postrema (AP) participates in the initiation of emetic reflexes elicited in various physiological, pathological or pharmacological circumstances (reviewed by Borison and Wang 1953; Wang 1965; Borison et al. 1981). In addition it is generally assumed that a large number of emetic and anti-emetic agents exert their actions in humans, at least in some cases, by interfering with local neurotransmitter metabolism or with receptors implicated in emetic reflexes (Peroutka and Snyder 1982). Nevertheless there are still a limited number of compounds used to control nausea and vomiting in patients; their efficacy is sometimes limited and their modes of loci of actions not always perfectly understood. Animal experiments are useful for such purposes but, in view of the known species differences in emetic reflexes, we found it of interest to study directly in human AP the presence of various neurotransmitters, their synthesising and inactivating enzymes and their receptors. This neurochemical investigation was performed on the AP dissected out from recently deceased patients and attention was particularly focused on markers of neurotransmitter systems likely to be involved in nausea and vomiting, as suggested by various drug effects, like the cholinergic, opioidergic, histaminergic and catecholaminergic systems. Indeed, one limitation in these kind of studies are the post-mortem alterations which may occur during the time elapsed between death and obtaining the tissue preparations used in the biochemical studies. In order to circumvent to a certain extent this inherent difficulty, we have, in most cases, studied the post-mortem stability of the various markers in the brains of laboratory animals or referred to previous studies on this matter.

Human Brain Samples

Human brains were obtained from 30–40 subjects (80–90 years) with no known neurological disease. The time elapsed between death and autopsy was 4–20 h. Within less than 2 h after autopsy, the brains were stored at −70 °C. The caudate nucleus was dissected out from tissue slices on a plate cooled to −15 °C (Javoy-

1 Unité 109 de Neurobiologie, Centre Paul Broca de l'INSERM, 2 ter rue d'Alésia, 75014 Paris, France
2 Laboratoire de Médecine Expérimentale, CHU Pitié Salpêtrière, Paris, France
3 Laboratoires de Physiologie, Université René Descartes, Paris, France

Nausea and Vomiting: Mechanisms and Treatment
Ed. by C. J. Davis, G. V. Lake-Bakaar and D. G. Grahame-Smith

Agid and Agid 1980). The AP was dissected out from brain stem by scraping the little pink prominence, 3–4 mm deep, localised bilaterally on the floor of the fourth ventricle just rostral to the obex. The various samples were frozen and stored at −70 °C until assay.

Cholinergic and GABA-ergic Markers

Choline acetyltransferase activity is considered as the most reliable marker for a cholinergic innervation. The enzyme activity in human AP was about one-fifth that of nucleus caudatus, the richest brain area (Table 1). This result can be considered reliable inasmuch as the enzyme activity in rat brain displays a very good

Table 1. Cholinergic and GABAergic markers

Markers	Area postrema	Nucleus caudatus
Choline acetyltransferase (nmol/h/mg protein)	12.5	62±9
^{3}H-Quinuclidyl benzylate binding (fmol/mg protein)	99.0	1,097±41
Glutamic acid decarboxylase (nmol/h/mg protein)	19	10

Means of triplicate determinations on pooled APs and means (±SEM) of 12 nucleus caudatus

Notes on method. For enzymatic activity determination, tissues were sonicated in 10 vol. of ice-cold 0.01 *M* Tris-acetate buffer pH 6 containing 0.2% Triton X-100. The homogenates were centrifuged ($8 \cdot 10^4$ $g \times$ min) and the supernatants stored at −70 °C until assay.
Choline acetyltransferase activity was determined according to the method of Fonnum (1969) as slightly modified (Puymirat et al. 1979). A 10 µl aliquot of tissue homogenate supernatant was incubated for 4 min at 37 °C with 12 µl of a mixture containing 0.2 *M* sodium phosphate buffer pH 7.4, 0.6 *M* NaCl, 0.05 m*M* [L-^{14}C]-acetyl CoA and 10 m*M* choline chloride. Incubations were stopped by adding 100 µl ethylbutylacetone-tetraphenylborate (10 mg/ml). After shaking and centrifugation, 50 µl of the organic upper layer were withdrawn and counted.

Glutamic acid decarboxylase activity was measured according to Puymirat et al. (1979). A 10 µl aliquot of the supernatant was incubated (30 min at 37 °C) with 27 µl of a mixture containing 10 m*M* sodium phosphate buffer pH 7, 10 m*M* β-mercaptoethanol, 0.02 m*M* pyridoxal phosphate and [U-^{14}C]-L-glutamic acid (3.5 m*M*, 0.19 µCi). Incubation tubes contained a strip of 3MM Whatman chromatographic paper spotted with 10 µl 1 *M* hyamine hydroxide. Reactions were stopped by injection of 100 µl 2M H_2SO_4 and further incubated for 15 min at 70 °C. Absorbed $^{14}CO_2$ on the hyamine hydroxide-spotted strip was then counted.
For *binding experiments,* homogenates were prepared by rapid sonication in ice-cold 50 m*M* Tris-HCl buffer pH 7.4 containing 120 m*M* NaCl, 5 m*M* KCl, 1 m*M* $CaCl_2$, 1 m*M* $MgCl_2$ ("Tris-ions" buffer). After centrifugation ($3 \cdot 10^5$ $g \times$ min) the particulate fraction was resuspended, incubated for 15 min at 37 °C and centrifuged 3 times. Before use, the pellet was resuspended in "Tris-ions" buffer (0.05 mg protein/incubation) and incubated for 60 min at 22 °C with 4 n*M* [^{3}H]-L-quinuclidinylbenzilate (^{3}H-QNB) in the absence or in the presence of 1 µ*M* atropine to determine the non-specific binding. The binding reaction was stopped by rapid filtration under reduced pressure through glass fibre filters Whatman GF/B.

post-mortem stability even at 20 °C (Puymirat et al. 1979). A relatively high level of ^{3}H-quinuclidyl benzylate binding, a marker of muscarinic receptors in human brain (Wastek and Yamamura 1978; Ruberg et al. 1982), was also found in human AP, representing about one-tenth of the binding in n. caudatus (Table 1).

These observations suggest that the AP may contain cholinergic neurons or receive a cholinergic input, the origin of which does not appear to have been as yet established by immunohistochemical studies with choline acetyltransferase antibodies. The latter technique has however been used to detect cholinergic neurons spreading from the facial nucleus through the reticular formation towards the nucleus solitarius (Cuello and Sofroniew 1984). In an autoradiographic study of rat brain stem, muscarinic receptors were localised in the nucleus tractus solitarius and nucleus ambiguus, but apparently not in the AP (Wamsley et al. 1981).

The presence of muscarinic receptors in the human AP is in agreement with the increase in firing frequency of spontaneously discharging cells (in explants) of rat AP under the action of carbamylcholine which is blocked by atropine (Brooks et al. 1983). Muscarinic antagonists like scopolamine appear to display clear anti-emetic capacity, but it remains doubtful whether they exert these effects at the level of AP because (a) brain-penetrating agents are generally used for this purpose and (b) high densities of muscarinic receptors (and of cholinergic innervation) are present in various other brain areas (like the nucleus tractus solitarius) involved in the emetic reflex.

A high level of glutamic acid decarboxylase activity, a reliable marker of GABA-ergic neurons, is also present in the human AP (Table 1). This is in good agreement with the observation of GABA levels in the rat AP which were among the highest of any brain-stem area in this species (Dietrich et al. 1982).

Histaminergic Markers

The enzyme L-histidine decarboxylase which is responsible for histamine biosynthesis represents the most reliable marker for histaminergic neurons in brain (reviewed by Schwartz 1975 and Schwartz et al. 1979). It has been previously characterised in human brain owing to its reasonable post-mortem stability (Barbin et al. 1980).

An easily measurable L-histidine decarboxylase activity was found in human AP, where its level was similar to that found in n. caudatus (Table 2), i.e. corresponding approximately to the mean cerebral level (Barbin et al. 1980). High histamine levels seem also to be present in AP (Adam 1961). Histamine synthesis in brain occurs not only in neurons but also in mast cells. The latter, mainly associated with perivascular structures, have been shown by histological techniques in the monkey AP (Brizzee et al. 1978). However, the histamine turnover in mast cells being extremely slow, these cells are characterised by a very low L-histidine decarboxylase activity, which may suggest the presence of histaminergic neurons (in which the synthesis rate is much higher) in the AP. Lesion studies have previously suggested the existence of a descending histaminergic neuronal pathway projecting to various brain-stem nuclei (Pollard et al. 1978). Whether it partici-

Table 2. Histaminergic markers in human brain regions

Markers	Area postrema	Nucleus caudatus
L-Histidine decarboxylase activity (dpm/h/μg protein)	1.9±0.1	1.8±0.2
^{3}H-Mepyramine binding (fmol/mg protein)	2.6±0.6	

Means (±SEM) of five separate experiments (triplicate)

Notes on method. Histamine H_1 receptor binding was measured in particulate fractions from area postrema (0.25 mg protein) prepared as described for rat or guinea-pig brain areas (see Table 3).

L-*Histidine decarboxylase activity* was measured in aliquots of the first homogenate obtained during membrane preparations, using the radiochromatographic assay of Baudry et al. (1973) with minor modifications. Homogenates (0.1 mg protein in 50 μl) were incubated with 3×10^6 dpm L-2,5[^{3}H]-histidine in a K_2/K-phosphate buffer 0.4 *M* pH 7.4 containing 10^{-5} *M* pyridoxal phosphate and 10^{-4} *M* histamine dihydrochloride. The synthesised ^{3}H-histamine was isolated using ion-exchange chromatography on amberlite CG50 columns (Garbarg et al. 1983).

pates in the innervation of the AP remains to be confirmed by one of the immunohistochemical techniques for either L-histidine decarboxylase (Watanabe et al. 1984; Pollard et al. 1985) or histamine (Panula et al. 1984; Steinbusch and Mulder 1984) with which histaminergic pathways in the CNS are currently being mapped out.

Among the various classes of cerebral histamine receptors, i.e. H_1, H_2, and H_3 characterised in mammalian brain (Schwartz et al. 1982a; Arrang et al. 1983) only H_1 receptors can be, at the present time, reliably studied with radioligand binding techniques, using ^{3}H-mepyramine (Hill et al. 1978; Chang et al. 1978). ^{3}H-mepyramine binding sites appear to display a good post-mortem stability: the number of these sites in guinea-pig cerebellum, one of the richest known tissues, did not significantly change up to 24 h following death (Table 3).

However, ^{3}H-mepyramine binding was hardly detectable in the human AP (Table 2). This is in agreement with an autoradiographic study of H_1-receptor localisation in the rat brain. Whereas high levels were present in neighbouring areas involved in the control of emetic reflexes like the nucleus tractus solitarius and the dorsal motor nucleus of the vagus, the AP was apparently devoid of H_1 receptors (Palacios et al. 1981). On the other hand following direct application of histamine onto the dog AP the emetic response which follows is partially blocked by H_1 antihistamines (Bhargava et al. 1976). It may well be that these effects are mediated by H_1 receptors in the neighbouring brain-stem areas and this may also be the case for the beneficial effects of H_1 antihistamines used in treating motion sickness. Indeed most agents used for this purpose easily cross the blood-brain barrier and occupy H_1 receptors in large brain areas (Quach et al. 1979). It would therefore be of interest to assess whether the more recent non-sedating H_1-receptor blockers with a preferential action at peripheral sites (Quach et al. 1980; Rose et al. 1982) also display any anti-nausea activity.

Table 3. Influence of post-mortem delay on histamine H_1, opiate and dopamine D-2 receptor stability

Post-mortem delay	Specific binding (fmol/mg protein)		
	^{3}H-Mepyramine	^{3}H-(D-Ala^2-Met^5) Enkephalinamide	^{3}H-Domperidone
2 min	134±6	365±15	270±22
1 h	140±9	374±33	254± 9
6 h	125±5	368± 5	286± 8
24 h	122±5	368±35	243±14

Notes on method. Guinea-pigs (for histamine H_1 receptors) or rats (for opiate and dopamine receptors) were killed by stunning, decapitated and the heads were left for various time intervals at ambient temperature. Cerebellum or striata were dissected out and frozen at −70 °C until assay. For preparation of the particulate fractions used in binding studies, brain regions were homogenised in 10 vol. of cold 50 m*M* Tris-HCl buffer, pH 7.4 and centrifuged twice (10^3 $g \times$ min and 2×10^5 $g \times$ min). The pellets were resuspended, incubated for 12 min at 37 °C and further centrifuged (2×10^5 $g \times$ min). The final particulate fractions were then resuspended in 10–15 vol. of the appropriate buffer and immediately used for the different binding studies.

Histamine H_1 receptors were quantified in aliquots of the guinea-pig cerebellar particulate fraction in 50 m*M* Tris HCl buffer, pH 7.4, containing 120 m*M* NaCl, 5 m*M* KCl, 1 m*M* $CaCl_2$, 1 m*M* $MgCl_2$, 0.1% ascorbic acid ("Tris-ions" buffer) by measuring the saturable binding of 2 n*M* ^{3}H-mepyramine, using 2.10^{-7} *M* mianserin to evaluate the non-specific binding; 450 µl freshly prepared tissue suspension (0.4 mg proteins) were preincubated for 20 min at 30 °C. Incubations initiated by addition of 50 µl ^{3}-mepyramine in the absence or presence of mianserin, an H_1-receptor antagonist (Garbarg et al. 1983), were carried on for 30 min and stopped by dilution into cold Tris buffer and rapid filtration (GF/B Whatman filters rinsed with 2×15 ml buffer).

Opiate receptors were quantified in aliquots of the particulate fraction from rat striatum resuspended in cold 50 m*M* Tris-HCl buffer pH 7.4, by measuring the binding of 15 n*M* ^{3}H(D-Ala^2-Met^5)enkephalinamide alone or in the presence of 5 µ*M* morphine for evaluation of the non-specific binding; 140 µl freshly prepared tissue suspension (0.15 mg protein) were preincubated and ligands added (under 60 µl) as previously described (Llorens et al. 1982). Incubations were stopped as for other bindings.

Dopamine receptors were quantified in rat striatal particulate fraction resuspended in "Tris-ions" buffer and incubated for 30 min at 30 °C with 4 n*M* ^{3}H-domperidone, in the absence or the presence of 50 µ*M* apomorphine to evaluate the non-specific binding. Incubations were stopped by rapid filtration, as already described (Baudry et al. 1979).

Markers for Opioids and Opioid Function

A relatively good post-mortem stability was found for the three selected opioidergic markers, i.e. (Met^5) enkephalin levels (Emson et al. 1980), opiate receptors with (D-Ala^2, Met^5) enkephalinamide and "enkephalinase" activity (Llorens et al. 1982) (Table 3).

Extremely high levels of (Met^5) enkephalin and opiate receptors seem to be present in the AP (Table 4) which therefore appears as one of the richest brain

Table 4. Opioid markers in human brain regions

Markers	Area postrema	Nucleus caudatus
(Met^5) enkephalin level (ng/mg protein)	(24.0)	12.3±2.7
^3H-(D-Ala^2-Met^5-NH_2) enkephalin binding (fmol/mg protein)	128±29	85±8
"Enkephalinase" activity (fmol/min/mg protein)	(6–32)	79±7

Means (±SEM) of five separate determinations (triplicate)

Notes on method. Endogenous Met^5-enkephalin levels were measured in aliquots of 0.1 *M* HCl, 0.005 *M* thioglycolic acid homogenate using a radioimmunoassay (Taquet et al. 1982).

Opiate receptors were quantified in particulate fractions prepared as described for rat brain areas in Table 3.

"Enkephalinase activity" was evaluated as described by Llorens et al. (1982) in the same particulate fraction using 40 n*M* (D-Ala^2-Leu^5-)-^3H-enkephalin as a substrate in the presence of 100 μ*M* puromycin, an aminopeptidase inhibitor (Barclay and Phillipps 1978) and 1 μ*M* captopril, an angiotensin converting enzyme inhibitor (Cushman et al. 1977). Blanks were obtained in the presence of 0.1 μ*M* Thiorphan, a selective "enkephalinase" inhibitor (Roques et al. 1980).

areas for these markers. High levels of (Met^5) enkephalin have also been reported by others (Grämsch et al. 1979); whereas, the other opioid peptide dynorphin seems much less abundant (Grämsch et al. 1982). The presence of opiate receptors in the AP is consistent with observations showing that enkephalins or morphine elicit emesis in dogs after either local (Bhargava et al. 1981) or systemic administration (Borison 1959; Carpenter et al. 1984), the latter responses being prevented by ablation of the AP.

A rather moderate level of "enkephalinase" (EC 3.4.24.11) activity was also present in the human AP (Table 4) as compared with the nucleus caudatus, one of the richest brain areas (Llorens et al. 1982). This membrane-bound enzyme, together with aminopeptidases, seems implicated in the physiological inactivation of enkephalins following their release (reviewed by Schwartz et al. 1981, 1982b; Schwartz 1983). It has even been hypothesised that its inhibition by strongly emetic compounds like cis-platinum could be indirectly responsible for nausea and vomiting in patients so treated (Harris 1982). However, this seems unlikely since (a) cis-platinum appears devoid of any significant "enkephalinase" inhibitory activity in vitro and (b) Thiorphan, a potent "enkephalinase" inhibitor (Roques et al. 1980), does not induce emesis in either dogs or humans (unpublished observations).

In spite of the fact that several markers of opioidergic synapses seem present in the AP and that exogenous opioid peptides display emetic activity, a physiological role for the endogenous peptides in emetic reflexes (that could have been demonstrated using antagonists like naloxone) apparently remains to be established.

Catecholaminergic Markers

The human AP contains measurable levels of the three catecholamines noradrenaline, adrenaline and dopamine (Table 5). These observations are in good agreement with previous reports showing a high level of noradrenaline in dog (Vogt 1954) or human (Farley and Hornykiewicz 1977) as well as with histochemical studies indicating high levels of catecholamines (Nobin and Bjorklund 1973) and of tyrosine-hydroxylase and dopamine-β-hydroxylase immunoreactivity in this area (Armstrong et al. 1981, 1982). In the rat, this seems to correspond to a high density of noradrenaline in the perikarya, proximal processes and varicose axon terminals which appear to form reciprocal connections between the AP and the medial nucleus of the tractus solitarius (Armstrong et al. 1981).

The presence of adrenaline in the human AP (Table 5) is consistent with the immunocytochemical detection of phenylethanolamine-*N*-methyltransferase in terminals throughout the rat AP which might emanate from perikarya in the adjacent medial nucleus of the tractus solitarius (Armstrong et al. 1981, 1982).

The significance of the relatively high level of dopamine (Table 5) is less clear since it may correspond to the catecholamine playing a precursor role in noradrenergic or adrenergic neurons as well as to the presence of dopaminergic neurons. The immunohistochemical studies of Armstrong et al. (1982) in rat AP, have failed to detect cells containing tyrosine hydroxylase without dopamine β-hydroxylase (like those present in the dorsal motor nucleus of the vagus). However, this does not rule out the possible occurrence of a dopaminergic innervation of the human AP obscured by the large noradrenergic and adrenergic innervations (see Lindvall and Bjorklund, 1983 for a discussion of this problem).

Table 5. Catecholaminergic markers in human brain regions

Markers	Area postrema	Nucleus caudatus
Noradrenaline (ng/mg protein)	4.0	4.6 ± 0.2
Dopamine (ng/mg protein)	1.2	35.4 ± 2.2
Adrenaline (ng/mg protein)	1.6	–
^{3}H-Domperidone binding (fmol/mg protein)	38 ± 7	167 ± 10

Means (±SEM) of four to six independent experiments

Notes on method. The *three catecholamine levels* were assayed using radio-enzymatic techniques (Da Prada and Zürcher 1976) on 250 μl aliquots of supernatants obtained after homogenisation of tissue in 40 vol. 0.2 *M* perchloric acid containing 7 m*M* $MgCl_2$ and 2.6 m*M* EGTA. Means of triplicate determinations on pooled APs and means (±SEM) of 12 n. caudatus. For *binding-site determination,* membranes from pooled APs (0.1–0.3 mg protein) or from nucleus caudatus (0.1 mg protein) were incubated with 4 n*M* ^{3}H-domperidone as already described in the notes in Table 3. Non-specific binding was determined in the presence of 50 μ*M* apomorphine.

The hypothesis of a dopaminergic innervation of the AP is reinforced by the observation that this structure contains a relatively high density of dopaminergic binding sites. The latter have been shown using either ^{3}H-domperidone (Table 5), a highly selective dopaminergic ligand (Baudry et al. 1979; Lazareno and Nahorski 1982), or (^{125}I)iodosulpiride (Table 6 and Fig. 1). It should be underlined that dopamine receptors assayed with ^{3}H-domperidone display a relatively good post-mortem stability (Table 3) and conservation in frozen membranes (results not shown).

At a nearly saturating concentration of ^{3}H-domperidone, the number of sites in human AP was about 20% of that found in nucleus caudatus, the richest human brain region. This relatively high density of dopaminergic sites was also observed in autoradiographic studies performed with (^{125}I)iodosulpiride (Fig. 1). On the same frontal sections not only the AP, but also neighbouring structures in the brain stem (some of which are known to be involved in the control of emetic reflexes), like the dorsal efferent nucleus of the vagus, the solitary nucleus, and the hypoglossal nucleus, were heavily labelled. The high selectivity of this new ligand (Martres et al. 1985 a, b) was shown by the complete prevention of labelling in the presence of 50 μ*M* apomorphine (compare Fig. 1 a and 1 b). This new ligand, because of its high specific radioactivity (2,000 Ci/mmol) is particularly well adapted to the assay of dopaminergic receptors in small brain areas. Thus detection of a significant specific binding was possible with (^{125}I)iodosulpiride at a concentration (0.4 n*M*) far below saturation and on membranes prepared from as little as 2–3 AP (Table 6).

Specific binding in this region was not inhibited by dopamine at 0.3 μ*M* a concentration close to the inhibition constant (Ki) value in the rat striatum (inhibition of 50% of the specific binding, Table 6). On the other hand, dopamine at 3 μ*M* inhibited by 50% the specific binding in the AP (indicating that the Ki value of dopamine is about 3 μ*M*). Thus, the affinity of dopamine for the human AP

Table 6. Some characteristics of ^{125}I-iodosulpiride binding

Conditions	Specific(^{125}I)iodosulpiride binding (fmol/mg protein)			
	Human area postrema		Rat striatum	
	−Gpp(NH)p	+Gpp(NH)p	−Gpp(NH)p	+Gpp(NH)p
Without dopamine	1.8±0.4	1.6±0.3	60±3	67±5
With 0.3 μ*M* dopamine	2.0±0.6	1.9±0.6	30±1[b]	45±3[a]
With 3 μ*M* dopamine	0.9±0.5	0.9±0.3	12±1[b]	23±2[b]

[a] $P<0.005$; [b] $P<0.001$ compared with respective total specific binding (without dopamine); means (±SEM) of two independent experiments performed in triplicate

Notes on method. Membranes (0.02–0.04 mg protein in "Tris-ions" buffer) were incubated with or without 100 μ*M* 5′-guanylylimidodiphosphate (GppNHp) and in the presence of 0.4 n*M* (^{125}I)iodosulpiride. Non-specific binding was determined in the presence of 25 μ*M* apomorphine. In addition dopamine at two concentrations was tested for its inhibition of the (^{125}I)iodosulpiride specific binding.

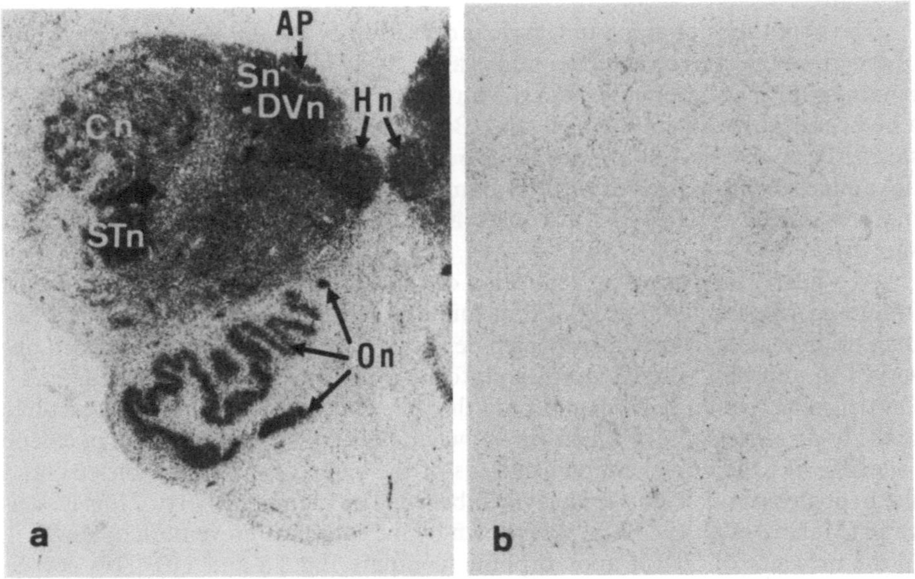

Fig. 1. Autoradiograms of human brain stem sections generated with (^{125}I)iodosulpiride. *Method.* Frontal cryostat sections (10 μm) were incubated for 30 min at 25 °C in "Tris-ions" buffer containing 0.3 n*M* ^{125}I-iodosulpiride without (**a**) or with (**b**) 50 μ*M* apomorphine to determine the non-specific binding. Following two 2 min rinses in ice-cold buffer, sections were dried and apposed on ^{3}H-ultrofilm (LKB) for 5 days. *AP*, area postrema; *Cn*, cuneate nucleus; *DVn*, dorsal efferent nucleus of vagus (X); *Hn*, hypoglossal nucleus (XII); *On*, olivary nucleus; *Sn*, solitary nucleus; *STn*, spinal trigeminal nucleus (V)

differs notably from that in the rat striatum, but is consistent with that reported by Stefanini and Clement-Cormier (1981) for the dog AP. Both ^{3}H-domperidone and (^{125}I)iodosulpiride display very low affinity for D-1 receptors which, therefore, are not labelled.

Guanine nucleotides have been shown to decrease the agonist affinity for the D-2 subclass of dopamine receptors (Creese et al. 1979; Sokoloff et al. 1980, 1984).

As already reported, the affinity of dopamine for the binding sites of the rat striatum was decreased by the guanyl nucleotide GppNHp (Table 6). Respective inhibitions by 0.3 and 3.0 μ*M* dopamine were significantly lowered in the presence of 100 μ*M* GppNHp, corresponding to a Ki value of about 1 μ*M* i.e. 3 times higher than in absence of GppNHp. In the same conditions, GppNHp did not affect the affinity of dopamine in the AP.

Dopamine receptors labelled with ^{3}H-spiperone were previously detected in the canine AP (Stefanini and Clement-Cormier 1981), but apparently displayed a higher sensitivity to guanyl nucleotides. It is not clear at present whether this represents a species difference in D-2 receptors or reflects post-mortem alterations in the human AP.

In any event, the present data clearly establish by using a biochemical approach, the occurrence of dopamine receptors in the human AP. This is in agreement with a large number of pharmacological observations that stimulation of dopamine receptors in this area triggers the emetic reflex and that D-2 dopamine receptor antagonists with a low brain penetrating ability, like domperidone or most of the benzamide derivatives, are potent anti-emetics (reviewed by Borison et al. 1981; Peroutka and Snyder 1982). However, the occurrence of dopamine receptors in other brain-stem areas involved in the control of the emetic reflex that is presently demonstrated (Fig. 1) suggests that dopamine antagonists which more easily cross the blood-brain barrier, i.e. neuroleptics, may have a larger anti-emetic efficacy. This would account for the fact that a compound like chlorpromazine inhibits the emesis induced in dogs by apomorphine, angiotensin II, neurotensin and enkephalins, whereas domperidone only blocks the response to apomorphine (Carpenter et al. 1984).

References

Adam HM (1961) Histamine in the central nervous system and hypophysis of the dog. In: Kety SS, Elkes J (eds) Regional neurochemistry. Pergamon, London, pp 293–306

Armstrong DM, Pickel VM, Joh TH, Reis DJ, Miller RJ (1981) Immunocytochemical localization of catecholamine synthesizing enzymes and neuropeptides in area postrema and medial nucleus tractus solitarius of rat brain. J Comp Neurol 196:505–517

Armstrong DM, Ross CA, Pickel VM, Joh TH, Reis DJ (1982) Distribution of dopamine-, noradrenaline-, and adrenaline-containing cell bodies in the rat medulla oblongata: demonstrated by the immunocytochemical localization of catecholamine biosynthetic enzymes. J Comp Neurol 212:173–187

Arrang JM, Garbarg M, Schwartz J-C (1983) Auto-inhibition of brain histamine release mediated by a novel class (H_3) of histamine receptor. Nature 302:1–5

Barbin G, Palacios JM, Garbarg M, Schwartz J-C, Gaspar P, Javoy-Agid F, Agid Y (1980) L-histidine decarboxylase in the human brain: properties and localization. J Neurochem 35:400–406

Barclay RK, Phillipps MA (1978) Inhibition of the enzymatic degradation of Leu-enkephalin by puromycin. Biochem Biophys Res Commun 81:1119–1123

Baudry M, Martres M-P, Schwartz J-C (1973) The subcellular localization of histidine decarboxylase in various regions of rat brain. J Neurochem 21:1301–1309

Baudry M, Martres M-P, Schwartz J-C (1979) ^{3}H-domperidone: a selective ligand for dopamine receptors. Naunyn Schmiedebergs Arch Pharmacol 308:231–237

Bhargava KP, Dixit KS, Palit G (1976) Nature of histamine receptors in the emetic chemoreceptor trigger zone. Br J Pharmacol 57:211–213

Bhargava KP, Dixit KS, Gupta YK (1981) Enkephalin receptors in the emetic chemoreceptor trigger zone of the dog. Br J Pharmacol 72:471–475

Borison HL (1959) Effect of ablation of medullary emetic chemoreceptor trigger zone on vomiting responses to cerebral intraventricular injection of adrenalin apomorphine and pilocarpine in the cat. J Physiol (Lond) 147:172–177

Borison HL (1983) A 1983 neuropharmacologic perspective of space sickness. Brain Behav Evol 23:7–13

Borison HL, Wang SC (1953) Physiology and pharmacology of vomiting. Pharmacol Rev 5:193–230

Borison HL, Borison R, McCarthy LE (1981) Phylogenic and neurologic aspects of the vomiting process. J Clin Pharmacol 21:23S–29S

Brizzee KR, Palazzo MC, Klara PM, Hofer H (1978) Effects of systemic administration of 6-hydroxydopamine on the circumventricular organs in non-human primates I. Area postrema. Cell Tissue Res 187:115–127

Brooks MJ, Hubbard JI, Sirett N (1983) Extracellular recording in rat area postrema in vitro and the effects of cholinergic drugs, serotonin and angiotensin II. Brain Res 261:85–90
Carpenter DO, Briggs DB, Strominger N (1984) Peptide-induced emesis in dogs. Behav Brain Res 11:277–281
Chang RSL, Tran VT, Snyder SH (1978) Histamine H_1-receptors in brain labeled with ^{3}H-mepyramine. Eur J Pharmacol 48:463–464
Creese I, Usdin T, Snyder SH (1979) Guanine nucleotides distinguish between two dopamine receptors. Nature 278:577–578
Cushman DW, Cheung HS, Sabo EF, Ondetti MA (1977) Design of potent competitive inhibitors of angiotensin-converting enzyme. Carboxylalkanoyl and mercaptoalkanoyl amino-acids. Biochemistry 16:5484–5491
Cuello AC, Sofroniew MV (1984) The anatomy of CNS cholinergic neurons. Trends Neurosci 7:74–78
Da Prada M, Zürcher G (1976) Simultaneous radioenzymatic determination of plasma and tissue adrenaline, noradrenaline and dopamine within the femtomole range. Life Sci 19:1161–1174
Dietrich WD, Lowry OH, Loewy AD (1982) The distribution of glutamate, GABA and aspartate in the nucleus tractus solitarius of the cat. Brain Res 237:254–260
Emson PC, Arregui A, Clement-Jones U, Sandberg BEB, Rossor M (1980) Regional distribution of methionine-enkephalin and substance P like immunoreactivity in normal human brain and in Huntington's disease. Brain Res 199:147–160
Farley IJ, Hornykiewicz O (1977) Noradrenaline distribution in subcortical areas of the human brain. Brain Res 126:53–62
Fonnum F (1969) Radiochemical microassays for the determination of ChAc and AChE activities. Biochem J 115:465–472
Garbarg M, Pollard H, Quach TT, Schwartz J-C (1983) Methods in brain histamine research. In: Parvez S, Nagatsu T, Nagatsu I, Parvez H (eds) Methods in biogenic amine research. Elsevier, Amsterdam, pp 623–662
Grämsch C, Höllt V, Mehraein P, Pasi A, Herz A (1979) Regional distribution of methionine enkephalin and β-endorphin like immunoreactivity in human brain and pituitary. Brain Res 171:261–270
Grämsch C, Höllt V, Pasi A, Mehraein P, Herz A (1982) Immunoreactive dynorphin in human brain and pituitary. Brain Res 233:65–74
Harris AL (1982) Cytotoxic therapy-induced vomiting is mediated via enkephalin pathways. Lancet 1:714–716
Hill SJ, Emson PC, Young JM (1978) The binding of ^{3}H-mepyramine to histamine H_1-receptors in guinea-pig brain. J Neurochem 31:997–1004
Javoy-Agid F, Agid Y (1980) Is the mesocortical dopaminergic system involved in Parkinson's disease? Neurology 30:1326–1330
Lazareno S, Nahorski SR (1982) Selective labelling of dopamine (D-2) receptors in rat striatum by ^{3}H-domperidone but not by ^{3}H-spiperidone. Eur J Pharmacol 81:273–285
Lindvall O, Björklund A (1983) Dopamine- and norepinephrine-containg neuron systems: their anatomy in the rat brain. In: Emson PC (ed) Chemical neuroanatomy. Raven, New York, pp 229–255
Llorens C, Malfroy B, Schwartz J-C, Gacel G, Roques BP, Roy J, Morgat JL, Javoy-Agid F, Agid Y (1982) Enkephalin dipeptidyl carboxypeptidases (enkephalinase) activity: selective radioassay, properties and regional distribution in human brain. J Neurochem 39:1081–1089
Martres M-P, Bouthenet M-L, Sales N, Sokoloff P, Schwartz J-C (1985a) Widespread distribution of brain dopamine receptors evidenced with ^{125}I-iodosulpride, a highly selective ligand. Science 228:752–755
Martres M-P, Sales N, Bouthenet M-L, Schwartz J-C (1985b) Localisation and pharmacological characterisation of D-2 dopamine receptors in rat cerebral neocortex and cerebellum using (^{125}I)sulpride. Eur J Pharmacol (in press)
Nobin A, Björklund A (1973) Topography of the monoamine neuron systems in the human brain as revealed in fetuses. Acta Physiol Scand [Suppl] 388:1–40
Palacios JM, Wamsley JK, Kuhar MJ (1981) The distribution of histamine H_1-receptors in the rat brain: an autoradiographic study. Neurosci 6:15–38

Panula P, Yang HYT, Costa E (1984) Histamine containing neurons in the rat hypothalamus. Proc Natl Acad Sci USA 81:2572–2576

Peroutka SJ, Snyder SH (1982) Antiemetics: neurotransmitter receptor binding predicts therapeutic actions. Lancet 2:658–659

Pollard H, Llorens C, Barbin G, Garbarg M, Schwartz J-C (1978) Histamine and histidine decarboxylase in brainstem nuclei: distribution and decrease after lesions. Brain Res 157:178–181

Pollard H, Pachot I, Legrain P, Buttin G, Schwartz J-C (1985) Development of a monoclonal antibody against L-histidine decarboxylase as a selective tool for the localisation of histamine-synthesizing cells. In: Ganellin CR, Schwartz J-C (eds) Frontiers in histamine research. Pergamon, New York, pp 103–117

Puymirat J, Javoy-Agid F, Gaspar P, Plaska A, Prochiantz A, Agid Y (1979) Post-mortem stability and storage in the cold of brain enzymes. J Neurochem 32:449–454

Quach TT, Duchemin AM, Rose C, Schwartz J-C (1979) In vivo occupation of cerebral histamine H_1-receptors evaluated with ^{3}H-mepyramine may predict sedative properties of psychotropic drugs. Eur J Pharmacol 60:390–392

Quach TT, Duchemin AM, Rose C, Schwartz J-C (1980) Labeling of histamine H_1-receptors in the brain of living mouse. Neurosci Lett 17:49–54

Roques BP, Fournie-Zaluski MC, Soroca E, Lecomte JM, Malfroy B, Llorens C, Schwartz J-C (1980) The enkephalinase inhibitor thiorphan shows antinociceptive activity in mice. Nature 288:286–288

Rose C, Quach TT, Llorens C, Schwartz J-C (1982) Relationship between occupation of cerebral H_1-receptors and sedative properties of antihistamines. Arzneimittelforsch 322:1171–1174

Ruberg M, Ploska A, Javoy-Agid F, Agid Y (1982) Muscarinic binding and choline acetyltransferase activity in Parkinsonian subjects with reference to dementia. Brain Res 232:129–139

Schwartz J-C (1975) Histamine as a transmitter in brain. Life Sci 17:503–518

Schwartz J-C (1983) Metabolism of enkephalins and the concept of neuropeptidase. Trends Neurosci 6:15–18

Schwartz J-C, Barbin G, Baudry M, Garbarg M, Martres MP, Pollard H, Verdiere M (1979) Metabolism and functions of histamine in the brain. In: Essman B, Valzelli L (eds) Current developments in psychopharmacology. Spectrum, New York, pp 173–261

Schwartz J-C, Malfroy B, De la Baume S (1981) Biological inactivation of enkephalins and the role of enkephalin dipeptidyl carboxypeptidase ("enkephalinase") as neuropeptidase. Life Sci 29:1715–1740

Schwartz J-C, Barbin G, Duchemin AM, Garbarg M, Llorens C, Pollard H, Quach TT, Rose C (1982a) Histamine receptors in the brain and their possible functions. In: Ganellin CR, Parsons ME (eds) Pharmacology of histamine receptors. Wright, Bristol, pp 351–391

Schwartz J-C, De la Baume S, Yi CC, Chaillet P, Marcais-Collado H, Costentin J (1982b) Enkephalin metabolism in brain and its inhibition. Prog Neuropsychopharmacol 6:665–671

Sokoloff P, Martres M-P, Schwartz J-C (1980) Three classes of dopamine receptors (D-2, D-3, D-4) identified by binding studies with ^{3}H-apomorphine and ^{3}H-domperidone. Naunyn Schmiedebergs Arch Pharmacol 315:89–102

Sokoloff P, Martres M-P, Delandre M, Redouane K, Schwartz J-C (1984) ^{3}H-domperidone binding sites differ in rat striatum and pituitary. Naunyn Schmiedebergs Arch Pharmacol 327:221–227

Stefanini E, Clement-Cormier YC (1981) Detection of dopamine receptors in the area postrema. Eur J Pharmacol 74:257–260

Steinbusch HWM, Mulder AH (1984) Immunohistochemical localization of histamine in neurons and mast cells in the rat brain. In: Björklund A, Hokfelt T (eds) The handbook of chemical neuroanatomy, vol 3, part II. Elsevier, Amsterdam, pp 126–140

Taquet H, Javoy-Agid F, Cesselin F, Hamon M, Legrand JC, Agid Y (1982) Microtopography of methionine enkephalin, dopamine and noradrenaline in the ventral mesencephalon of human control and Parkinsonian brains. Brain Res 235:303–314

Vogt M (1954) The concentration of sympathin in different parts of the central nervous system under normal conditions and after the administration of drugs. J Physiol (Lond) 123:451–481

Wamsley JK, Lewis MS, Young WS, Kuhar MJ (1981) Autoradiographic localisation of muscarinic cholinergic receptors in rat brainstem. J Neurosci 1:176–191

Wang SC (1965) Emetic and antiemetic drugs. In: Root WS, Hofman FG (eds) Physiological pharmacology. Academic, New York, p 255
Wastek GJ, Yamamura HI (1978) Biochemical characterization of the muscarinic cholinergic receptor in human brain: alterations in Huntington's disease. Mol Pharmacol 14:768–780
Watanabe T, Taguchi Y, Shiosaka S, Tanaka J, Kubota H, Terano Y, Tohyama M, Wada H (1984) Distribution of the histaminergic neuron system in the central nervous system of rats; a fluorescent immunohistochemical analysis with histidine decarboxylase as a marker. Brain Res 295:13–25

The Central Nervous Connections Involved in the Vomiting Reflex

K. R. Brizzee [1] and W. R. Mehler [2]

Introduction

The vomiting reflex may be elicited by a number of different types or classes of stimuli involving many varieties of receptor structures and considerable diversity in afferent pathways and central connections. Central relay or mediating structures thus may vary widely according to the type of initial emetic stimulus.

The emetic circuits which have been most completely delineated to date are probably those in which the chemoreceptor trigger zone (CTZ) in the area postrema (AP) functions as a key mediating structure. Even in this system, however, there are large gaps in our knowledge of the nerve tracts and central nervous connections involved.

Knowledge of most other emetic circuits subserving the emetic reflex resulting from many diverse types of stimuli such, for example, as emotional stress (e.g. psychogenic vomiting, Wruble et al. 1982), pain (e.g. testicular trauma), and chemical or mechanical irritation of the gastrointestinal tract or urinary tract is quite incomplete at this time, thus precluding any very adequate description of their central connections at present.

One physiological system, however, which has received considerable attention recently in relation to the vomiting reflex elicited by motion stimuli is the vestibular system. Due to the paucity of data on central nervous connections of several of the non-vestibular types of emetic stimuli cited above, we will devote most of our attention in this brief review to the central connections of the vestibular system which seem likely to be involved in the vomiting response to motion stimuli. However, the latter part of the review will be concerned with the concept of the reticular vomiting centre in relation to the parvicellular reticular formation (PCRF), and will thus probably pertain to all of the many classes of emetic stimuli since it will address the question of the final common emetic pathway.

Vestibular System Connections

Vestibular Receptors for Motion-Induced Vomiting

The vestibular receptor elements for motion-induced vomiting have been clearly shown to be located in the vestibular labyrinths, with motion-generated impulses

1 Department of Neurobiology, Delta Regional Primate Research Center, Tulane University, Three Rivers Road, Covington, Louisiana 70433, USA
2 NASA/Ames Research Center, Moffett Field, California 94035, USA

Nausea and Vomiting: Mechanisms and Treatment
Ed. by C. J. Davis, G. V. Lake-Bakaar and D. G. Grahame-Smith

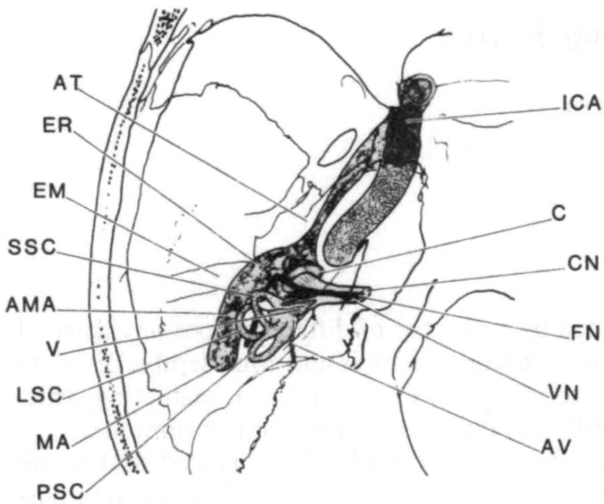

Fig. 1. Left auditory apparatus as if viewed through a semi-transparent temporal bone. (From Williams and Warwick 1975)

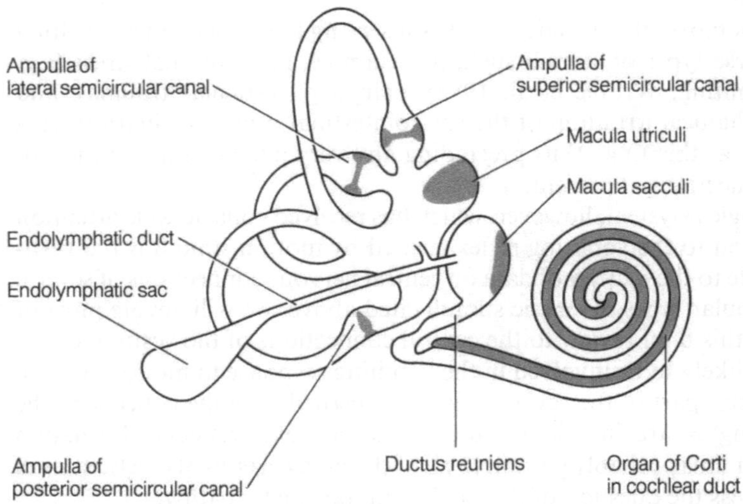

Fig. 2. Schema of the interconnections and spatial relationships of the semicircular canals, utricle and saccule together with the distribution of the eighth cranial nerve. (From Copenhaver et al. 1971)

conducted to the CNS via primary vestibular fibres in the eighth cranial nerve. The location and orientation of the labyrinths, including the semicircular canals, utricle and saccule in the petrous temporal bones are shown in Fig. 1. A schema of the interconnections and spatial relationships of the semicircular canals, utricle and saccule and the location of the receptor elements within them is shown in Fig. 2.

The actual receptor elements through which motion-emetic stimuli are generated appear to be the air cells in the cristae of the ampullae of the semicircular

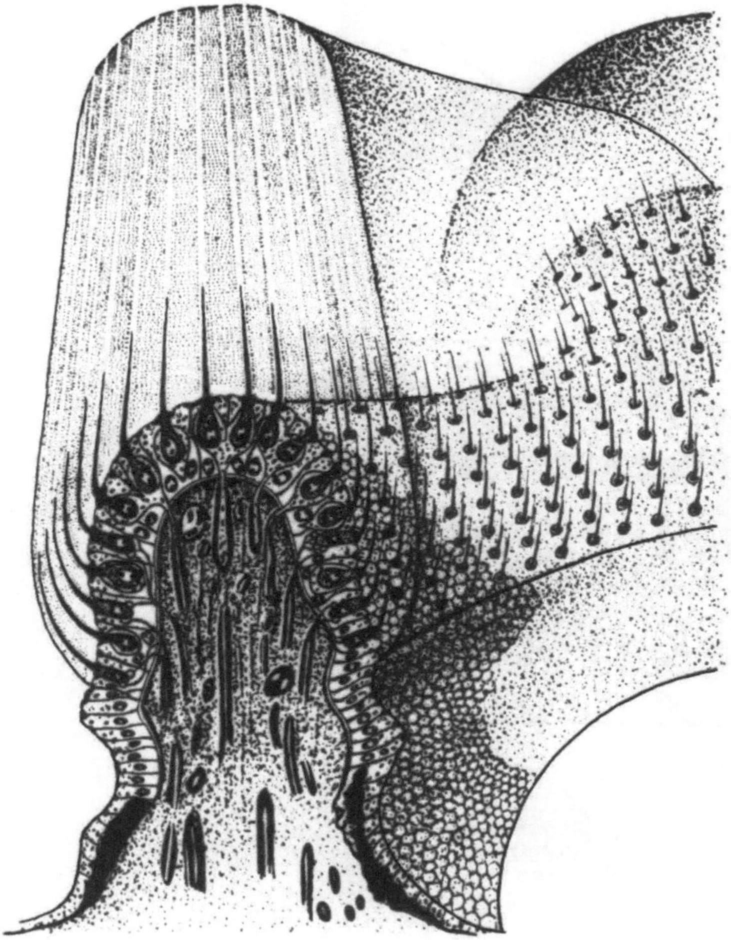

Fig. 3. Schematic drawing of one-half of a crista ampullaris, showing sensory epithelium, including receptor hair cells and supporting cells. (From Wersall 1966, p 1)

canals and also those in the maculae of the utricle and sacule (otolith organs). These latter structures are illustrated in Fig. 3 and 4.

The ultrastructural characteristics of the two types of hair cells as illustrated in Fig. 5 appear to be identical in the cristae ampullares and in the utricular and saccular maculae.

The actual data on which the conclusion as to the labyrinthine location of the receptors for motion-generated emesis is based have been derived from a large number of observations and studies on both human and animal subjects over a number of years. Using a modified elevator and crane device Sjoberg, (1931) found that labyrinthectomized dogs were completely insensitive to prolonged oscillation. Prior to the surgical interference, the same dogs had been readily "made sick" by similar motion. Brown et al. (1941) reported that patients whose laby-

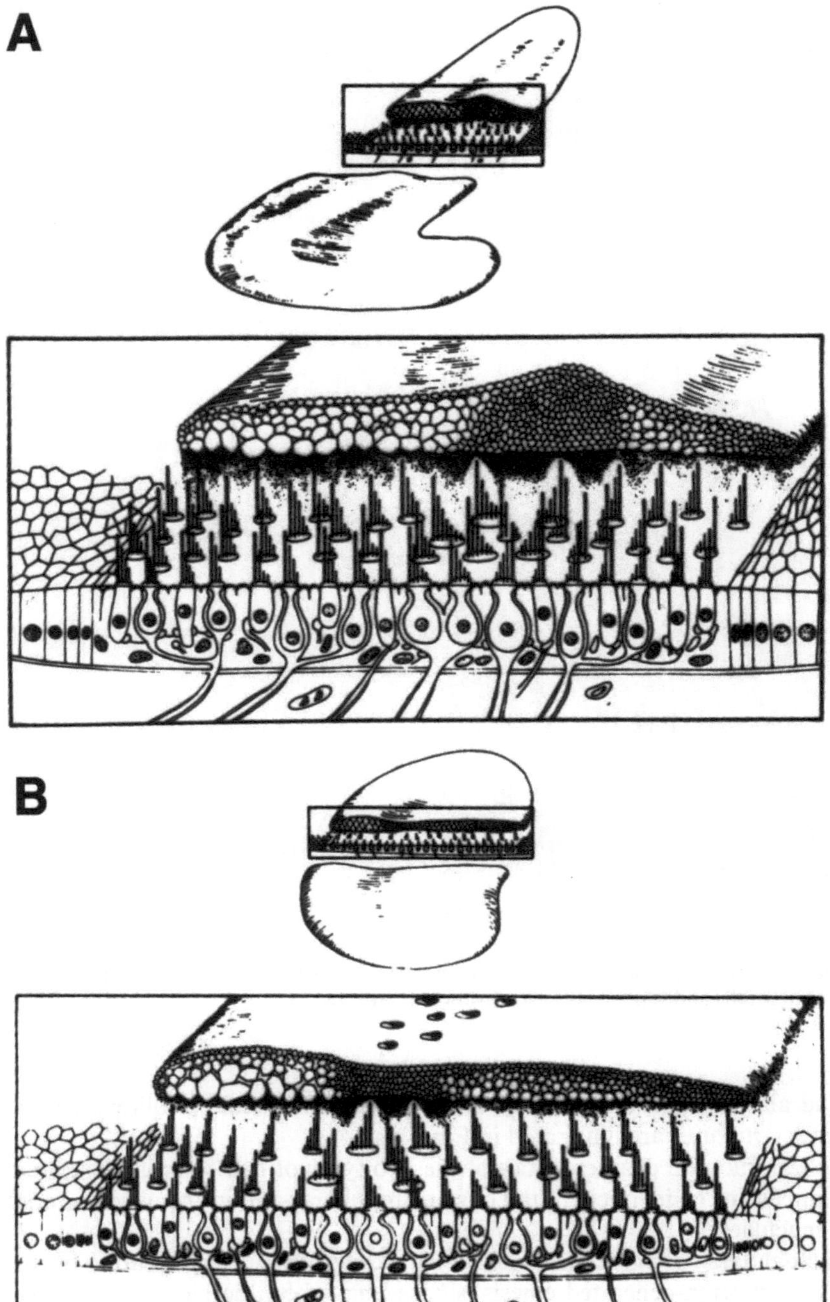

Fig. 4. Schematic drawings of (**A**) saccular and (**B**) utricular maculae showing hair cells and otolithic membrane of each. (From Lindemann 1973, p 411)

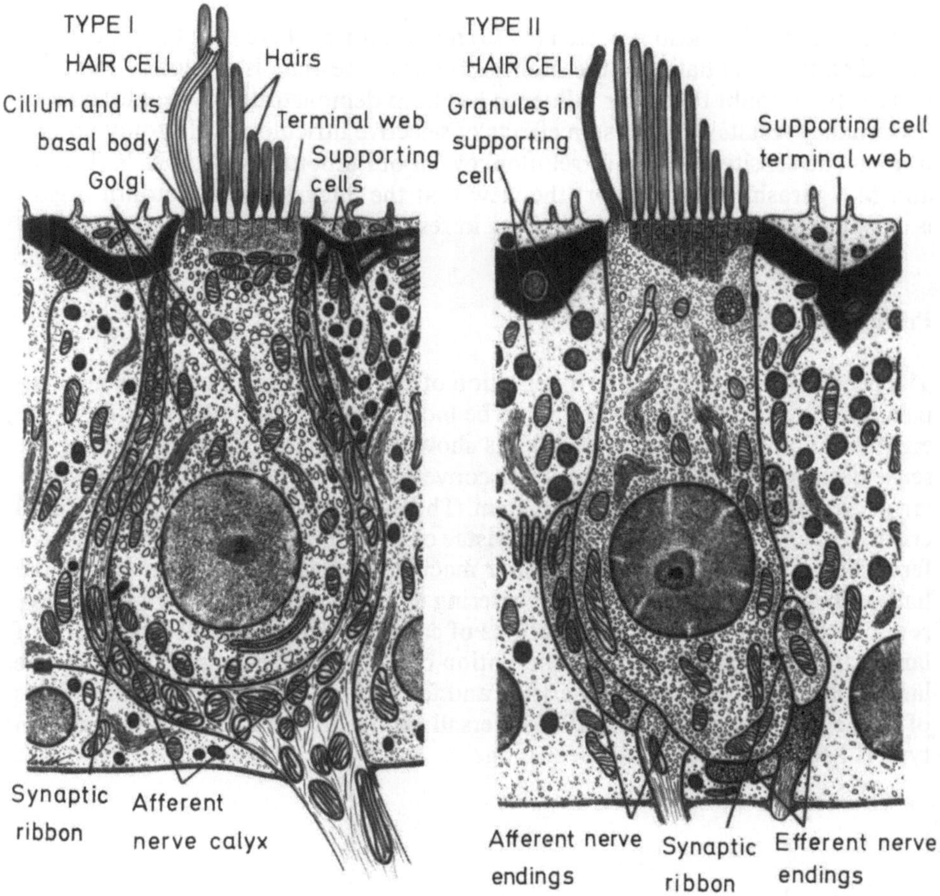

Fig. 5. Schematic representation of the principal ultrastructural features of the vestibular type I and type II hair cells and their supporting cells. (From Bloom and Fawcett 1975)

rinths had been destroyed by meningitis did not show any evidence of motion sickness during a swing test which normally elicited this phenomenon. Graybiel (1965) reported that labyrinthine-defective human patients, exposed to several motion parameters which provoke motion sickness in most normal individuals, did not experience reactions characteristic of motion sickness.

Several workers, including McNally et al. (1942), Babkin and Bornstein (1943), Wang and Chinn (1956) and Kurashvili (1964) showed that bilateral destruction of the vestibular receptors in dogs eliminated motion sickness in these animals. Similarly, Johnson and Taylor (1961) showed that bilateral labyrinthectomy eliminated motion sickness in susceptible squirrel monkeys and cats. Money and Friedberg (1964) demonstrated that inactivation of all six semicircular canals by plugging eliminates motion sickness in susceptible dogs to about the same degree as does bilateral labyrinthectomy in this species.

Miller and Graybiel (1972) showed that stimuli which are "adequately provocative and unique for the canals" evoked motion sickness in human subjects.

Graybiel and Miller also reported (1970) that although there may be some doubt regarding the essentiality of the otolith organs in the genesis of motion sickness, there is little doubt that their influence has been demonstrated in weightlessness. Preliminary results of studies on effects of selective utriculosacculectomy on motion-emetic sensitivity on squirrel monkeys in our laboratories (by K. R. Brizzee and M. Igarashi 1984) support the view that the integrity of the otolith organs is essential to the initiation of vomiting in response to motion.

Primary Vestibular Afferent Fibres

The nerve cells involved in the innervation of the hair cells of the semicircular canals, utricle and saccule are known to be located in the vestibular ganglia of the eighth cranial nerve. Gacek (1969) has shown that the afferents from all of the sense organs in the semicircular canals converge in the rostral half of the vestibular nerve before entering the brain stem. The afferents from the posterior canal crista are caudal to those serving the cristae of the lateral and superior canals. Afferents from the saccular and utricular maculae are located in about the caudal half of the vestibular nerve prior to entering the brain stem. Gacek's studies have revealed that there are two populations of canal afferents: a small population of large calibre fibres and a larger population of small-sized myelinated fibres. The large fibres enter the top of the cristae and terminate as large calyciform endings of type I hair cells (Spoendlin 1966; Wersall 1960). The small fibres terminate on type II hair cells in bouton-type endings.

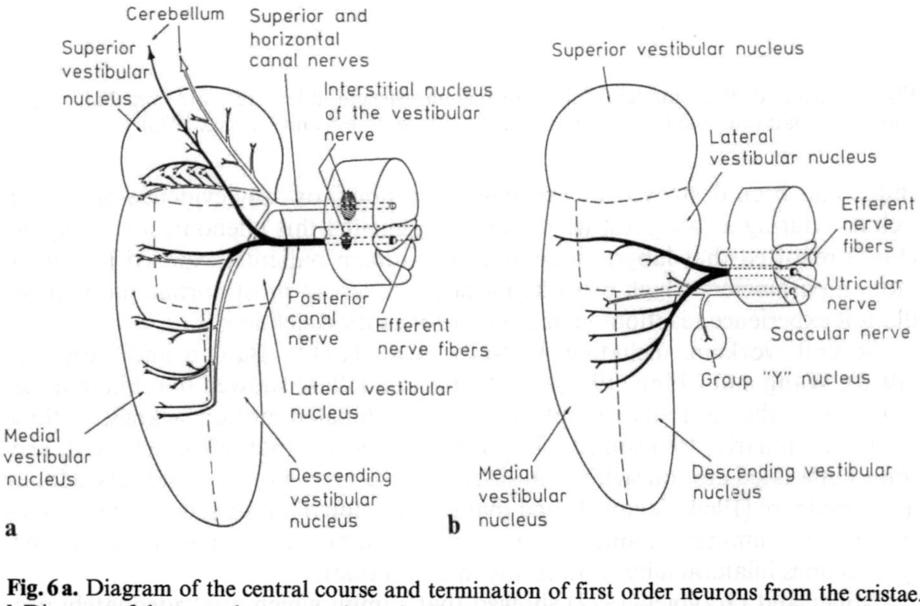

Fig. 6 a. Diagram of the central course and termination of first order neurons from the cristae. **b** Diagram of the central course and termination of first order neurons from the maculae. (From Gacek 1971, p 60)

The distribution of the primary afferent vestibular fibres in the cat was described in 1969 by Gacek (Fig. 6a, b). He noted that the primary canal afferents of the rostral half of the vestibular nerve give off numerous collaterals to the interstitial nucleus of the vestibular nerve just before they enter the brain stem. As the canal afferents enter the brain stem in the region of the lateral vestibular nucleus, they all bifurcate with an ascending branch passing through the superior vestibular nucleus in a dorsomedial direction. The posterior canal fibres terminate in the caudal and central part of the superior nucleus, and those from the superior caudal andn central part of the superior nucleus, and those from the superior division terminate more rostrally. All branches of the canal afferents give off long collaterals in the caudal portion of the superior nucleus which reach to cells in the ventromedial region of this nucleus. The ascending rami of the canal afferents then ascend into the vestibulocerebellum (Fig. 7). The smaller primary canal afferents terminate more laterally and ventrally in the superior nucleus and give off collaterals to the medial portion of the nucleus (Fig. 6a). The continuation of these afferents also ascend into the vestibulocerebellum.

The descending branches of primary canal afferents give off collaterals which penetrate the lateral vestibular nucleus and terminate on cells in the rostral part of the medial vestibular nucleus. These fibres give off collaterals to all parts of the medial nucleus.

Primary afferents from the utricular macula (Fig. 6b) bifurcate in the lateral vestibular nucleus. The ascending branch traverses the ventral division of the lat-

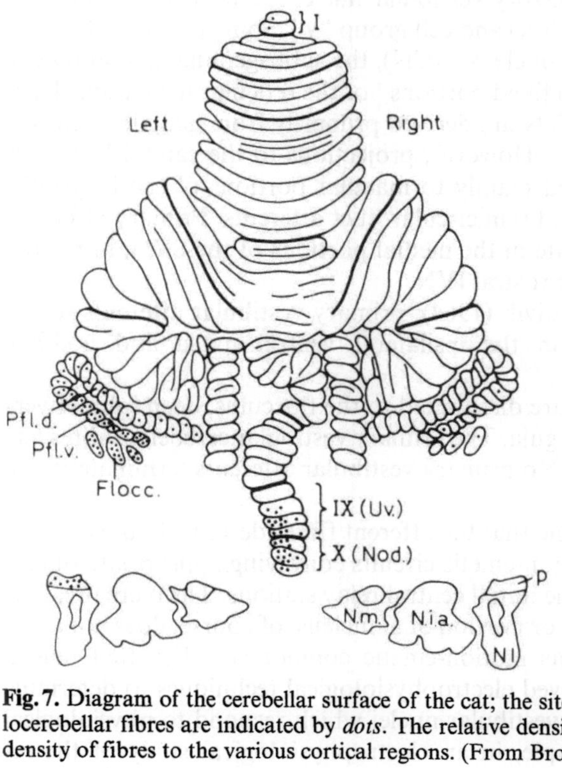

Fig. 7. Diagram of the cerebellar surface of the cat; the sites of termination of primary vestibulocerebellar fibres are indicated by *dots*. The relative densities of the dots give the approximate density of fibres to the various cortical regions. (From Brodal and Hoivik 1964, p 16)

eral nucleus, making contact with large neurons in this structure and medium-sized neurons in the medial nucleus. These branches do not continue into the cerebellum. The descending vestibular root makes contact with cells in the medial nucleus by collateral branches.

The primary saccular afferents reach the brain stem in the caudal edge of the vestibular nerve before bifurcating; the ascending branch then terminates rostrally in the Y nucleus, while the descending branch terminates in the medial vestibular nucleus and some parts of the lateral nucleus. The medial nucleus is the one area where collaterals of descending branches from all cristae and macular afferents converge on second-order vestibular neurons.

Carleton and Carpenter (1984), employing retrograde and anterograde axoplasmic transport techniques, recently studied the distribution of the primary vestibular fibres derived from the vestibular ganglion cells in the various vestibular nuclei in the squirrel monkey. They reported that the fibres to the superior (SVN) and medial (MVN) vestibular nuclei are distributed throughout these structures.

However, within SVN, terminals are most pronounced in its central large-celled portion, while the principal projections to the MVN are distributed evenly throughout the nucleus except for a small circular area of sparse terminals along its ventral margin. Primary vestibular afferents terminate mainly in rostral and caudal portions of the inferior vestibular nucleus (IVN). The primary afferents to the lateral vestibular nucleus (LVN) are restricted to its ventral part.

Primary afferents to the accessory vestibular nuclei reach the interstitial nucleus of the vestibular nerve (NIVN) and cell group “y”. Some primary afferents project to the accessory cuneate nucleus (ACN), the subtrigeminal lateral reticular nucleus (SLRN) and “well-defined portions” of the reticular formation (RF). The projections to SVN and MVN are derived primarily from ganglion cells innervating the semicircular ducts. However, projections to the caudal IVN, cell group “y” and ACN are related mainly to macular portions of the labyrinth. NIVN receives both macular and semicircular duct afferents. Primary afferents from the posterior duct terminate in the medial portions of the SVN, in rostrolateral portions on MVN and in rostral IVN.

According to Brodal and Hoivik (1964), primary vestibular afferents to the cerebellum terminate largely in the ipsilateral ventral uvula and nodulus (Fig. 7).

A smaller number of fibres are distributed to the flocculus, deep folia of vermal lobules V and VI and the lingula. The primary vestibular cerebellar fibres terminate as mossy fibre rosettes. No primary vestibular afferents terminate in the primate fastigial nuclei.

It seems reasonable to assume that the afferent fibres described above are all integral components of the motion-emetic circuits conveying appropriate signals from the receptor elements to the initial central relay stations. The mere presence of such fibres in a given nucleus or portion of a nucleus, of course, does not automatically identify such fibres as motion-emetic components. For that reason many investigators have employed electrophysiological techniques to determine the location of neurons within vestibular nuclei which respond to physiological or electrical stimuli. For example, in an electrophysiological study of the responses of 600 neurons in the vestibular nuclei to constant angular accelerations and decelerations of 4° per second, Ryu and McCabe (1973) classed vestibular neurons according to functional response type. Type I was an ipsilateral unit demonstrating an increased firing rate upon ampullopetal deflection of the ipsilateral cupula in the horizontal semicircular duct, and decreased firing rate upon ampullofugal deflection. Type II was a neuron showing a response opposite to that of type I. Type III showed decreased responses with either direction of cupula deflection, and type IV showed increased responses with either direction of cupula deflection. Type V showed no change in resting frequency of action potentials upon either direction of cupula deflection. According to these authors, the histological mapping of these neurons according to functional type tended to confirm the existing anatomical studies on the termination of primary vestibular fibres. Other authors who have carried out such vestibular mapping studies employing electrophysiological or natural stimulation have not clearly delineated the topographical relationships between their findings and the observations made in neuroanatomical studies.

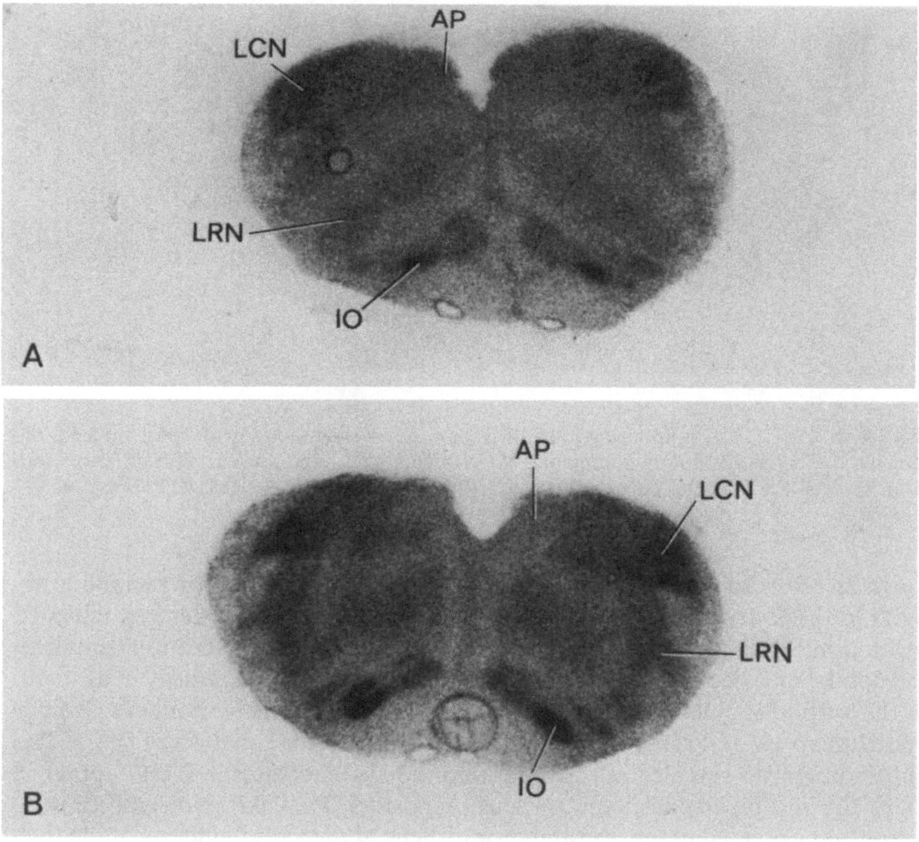

Fig. 8 A, B. [^{14}C]-2-DG autoradiographs of transverse cryostat sections of lower medulla oblongata. **A** Control. **B** Motion stimulated. *AP*, area postrema; *LCN*, lateral cuneate nucleus; *LRN*, lateral reticular nucleus; *IO*, inferior olive

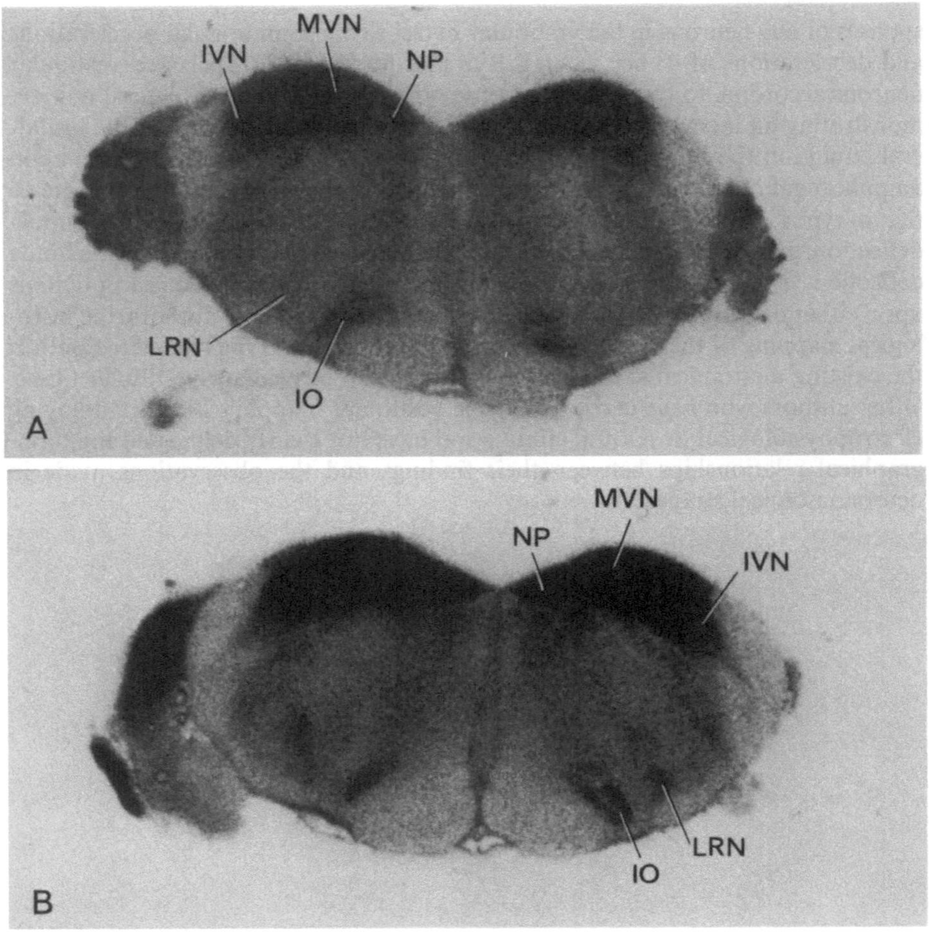

Fig. 9 A, B. [^{14}C]-2-DG autoradiographs of transverse cryostat sections of upper medulla oblongata. **A** Control. **B** Motion-stimulated. *MVN*, medial vestibular nucleus; *IVN*, inferior vestibular nucleus; *LCN*, lateral cuneate nucleus; *LRN*, lateral reticular nucleus; *IO*, inferior olive

In an effort to gain information on the influence of motion on metabolic activity in these structures, and to begin to map out the brain structures activated by motion stimuli, we have initiated studies on the influence of motion stimuli on the uptake of 2-deoxyglucose (2-DG) (Sokoloff et al. 1977; Kennedy et al. 1978) in the vestibular nuclei and other brain-stem sites in squirrel monkeys (Brizzee and Dunlap 1983). Preliminary quantitative densitometric analysis of [^{14}C]2-DG autoradiographs revealed selective motion-induced increases in 2-DG uptake in all of the vestibular nuclei and nucleus prepositus ($P < 0.05$). No significant increase was observed in the dorsal motor nucleus of the vagus, nucleus of the tractus solitarius or nucleus of the trigemino-spinal tract. The increases in 2-DG uptake in response to motion are illustrated autoradiographically (Figs. 8 A, B; 9 A, B; and 10 A, B). Quantitative densitometric analyses of 2-DG uptake in the above

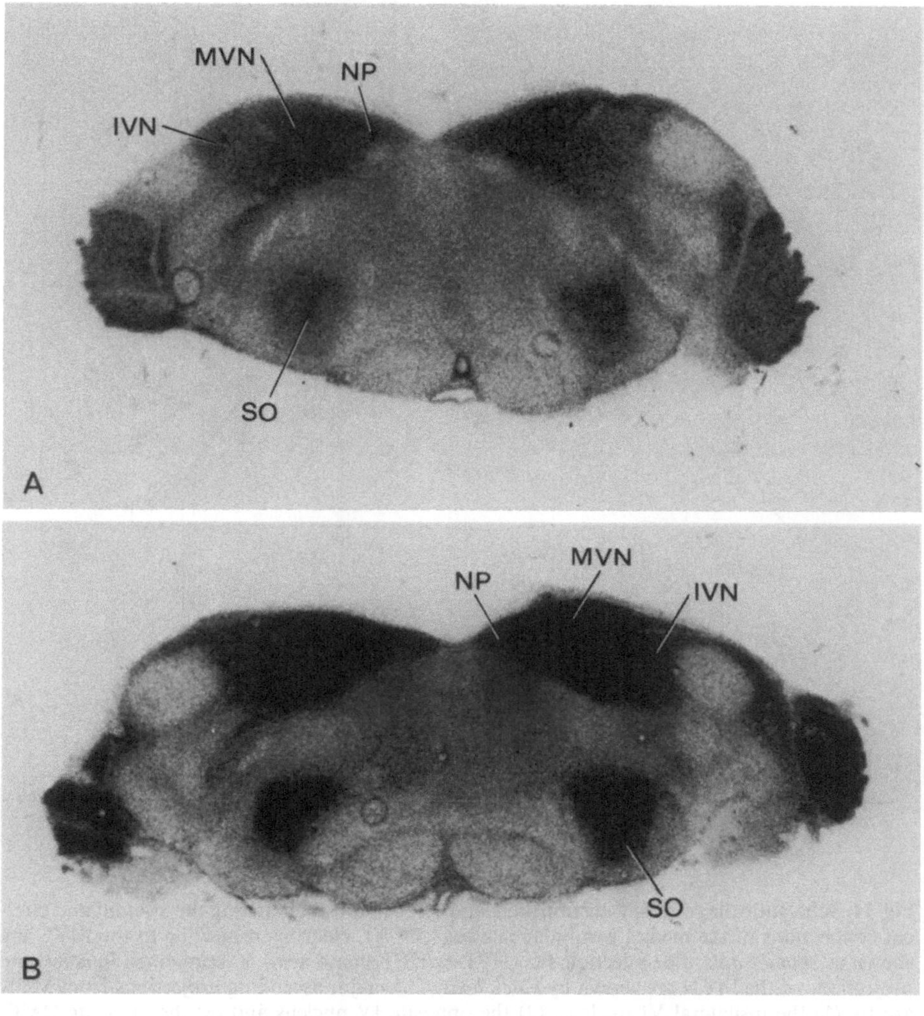

Fig. 10 A, B. [^{14}C]-DG autoradiographs of transverse cryostat sections of junction zone of medulla oblongata and pons. **A** Control. **B** Motion-stimulated. *NP*, nucleus prepositus; *MVN*, medial vestibular nucleus; *IVN*, inferior vestibular nucleus; *SO*, superior olive

nuclei and in several other nuclei and areas throughout the brain are in progress.

Secondary Vestibular Afferent Fibres

Studies to determine the secondary projections and connections of the vestibular nuclei and cerebellum in the squirrel monkey have been carried out by Carleton and Carpenter (1983) employing retrograde and anterograde axoplasmic transport techniques. The projections of secondary vestibular fibres from the MVN are

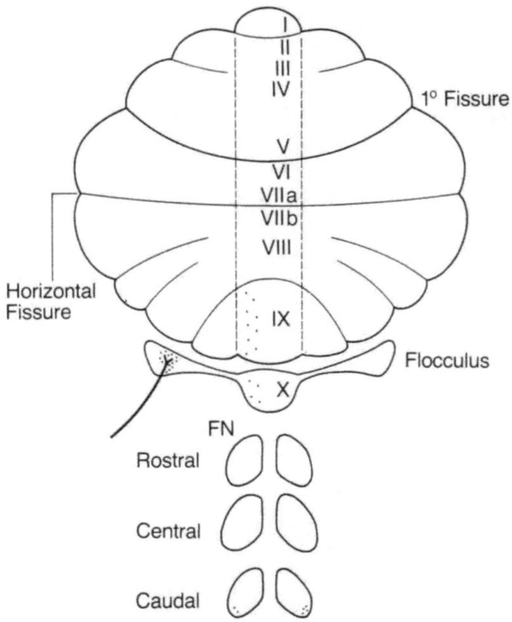

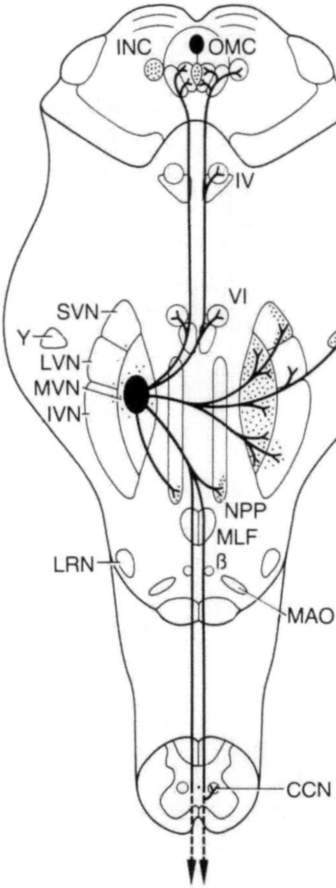

Fig. 11. Schematic diagram of the brain stem and cerebellum summarizing the afferent and efferent connections of the medial vestibular nucleus (*MVN*). Neurons projecting to the MVN are shown as *smaller dots*. The injection site (HRP or [^{3}H] amino acid) is represented in *black* and projections of the MVN are shown by *black lines*. The major ascending projections from MVN are to: (1) the ipsilateral VI nucleus, (2) the opposite IV nucleus and (3) the opposite OMC. (From Carleton and Carpenter 1983, p 32)

shown in Fig. 11, those of the IVN in Fig. 12 and those of the LVN in Fig. 13. The authors emphasize that the most impressive brain-stem afferents to the vestibular nuclei are commissural fibres with the MVN receiving and projecting the major commissural bundles. Fibres from the superior nucleus, as determined by several authors, enter the ipsilateral medial longitudinal fasciculus and ascend to terminate in the nuclei of cranial nerves III, IV and VI (Gacek 1971; Highstein 1971; Tarlov 1970). This nucleus also gives rise to a large complement of commissural fibres to the contralateral superior nucleus and medial nucleus (Gacek 1978). Projections from MVN are distributed bilaterally to the abducens nucleus (AN) and oculomotor complex (OMC) and contralaterally to the trochlear nucleus (TN). Ascending fibres from IVN terminate contralaterally only in the TN and parts of OMC.

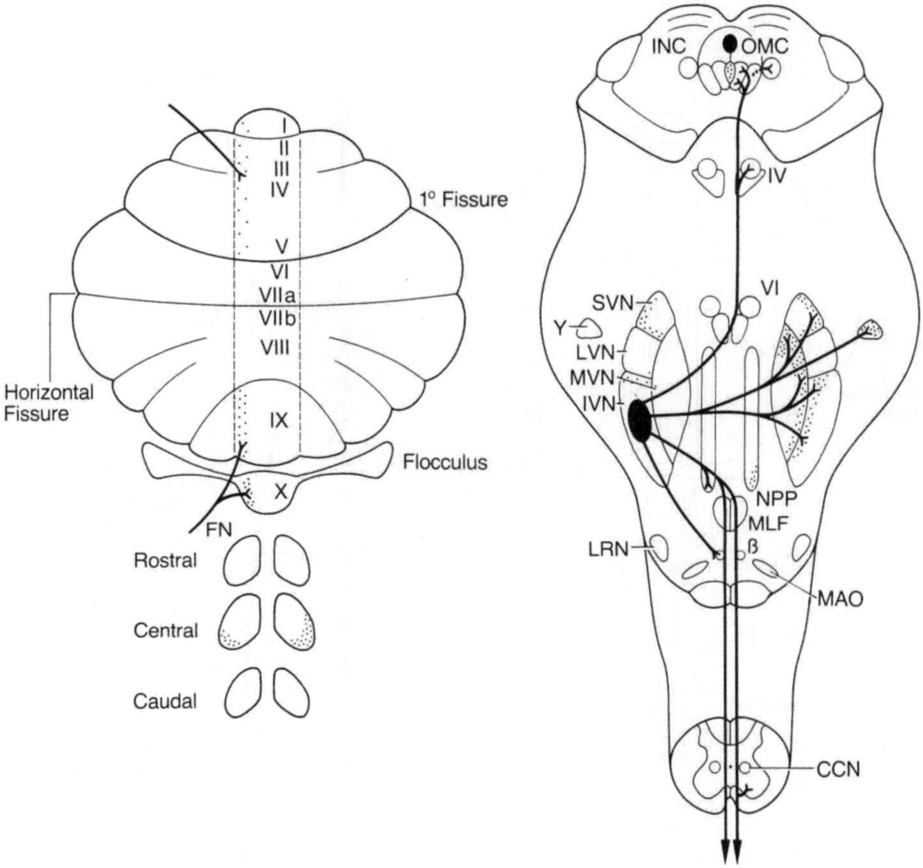

Fig. 12. Schematic diagram of the afferent and efferent projections of the inferior vestibular nucleus (*IVN*) in the brain stem and cerebellum. Neurons projecting to IVN are shown as *small dots*. The injection site (HRP or [^{3}H] amino acid) is represented in *black*, and projections from IVN are shown by *black lines*. Projections to the paramedian reticular nucleus are not shown. (From Carleton and Carpenter 1983, p 34)

Perihypoglossal Afferent Connections

The perihypoglossal nuclei (PH) are connected bilaterally and reciprocally with the IVN and MVN, and the latter receives fibres from the ipsilateral IVN and projects to the corresponding structure on the opposite side. The central cervical nucleus (CCN) is the only spinal nucleus projecting to the vestibular nuclei (Carleton and Carpenter 1983).

Efferent Projections from Nucleus Fastigius

In retrograde transport (HRP) studies on the connections of the fastigial nucleus in the squirrel monkey and the cat, Carpenter and Batton (1982) reported that

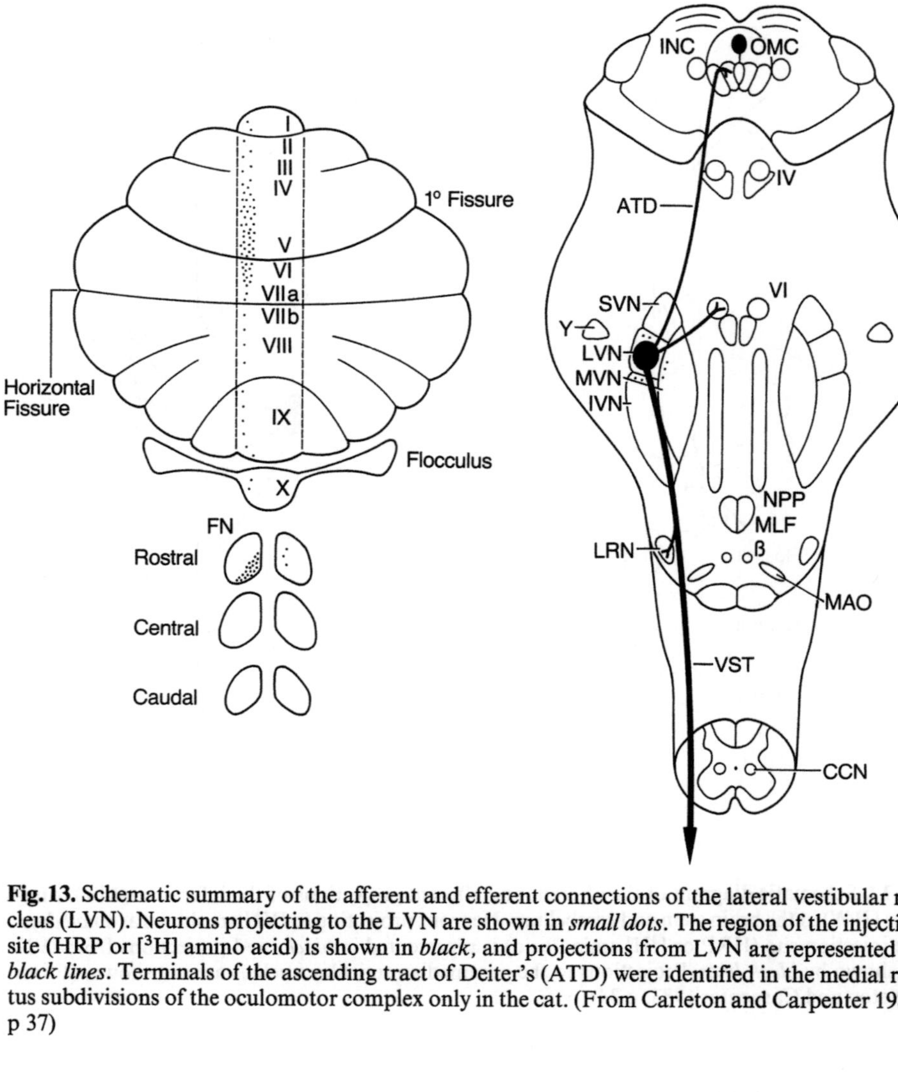

Fig. 13. Schematic summary of the afferent and efferent connections of the lateral vestibular nucleus (LVN). Neurons projecting to the LVN are shown in *small dots*. The region of the injection site (HRP or [^{3}H] amino acid) is shown in *black*, and projections from LVN are represented by *black lines*. Terminals of the ascending tract of Deiter's (ATD) were identified in the medial rectus subdivisions of the oculomotor complex only in the cat. (From Carleton and Carpenter 1983, p 37)

this nucleus receives afferents from Purkinje cells of the ipsilateral cerebellar vermis, nucleus B and the dorsomedial cell column of the contralateral inferior olive; the dorsal, lateral and caudal regions of the medial and inferior vestibular nuclei on both sides; cell group Z on both sides, and portions of the nucleus of Roller, nucleus propositus, and dorsal paramedian reticular nucleus.

Crossed efferent fibres in the uncinate fasciculus arise from all parts of the fastigial nucleus and outnumber those in the juxtarestiform body. These fibres project bilaterally to ventral and lateral portions of the IVN and LVN. No fastigial efferents terminate in the SVN or in the dorsal part of the LVN. Some crossed fastigial efferents descending in the IVN terminate in the nucleus parasolitarius (PSol).

Fastigioreticular fibres arise largely from the rostral portions of the fastigial nucleus, are mainly crossed and project to the medial regions of the nucleus reticularis gigantocellularis, parts of the caudal pontine reticular formation, the dorsal paramedian reticular nucleus, and portions of the magnocellular portion of the lateral reticular nucleus.

Descending fibres to the spinal cord are derived from the fastigiopontine bundle and course ventral to the trigeminospinal tract to the lateral funiculus of the spinal cord, extending as far as the upper cervical segments and apparently make monosynaptic contact with motor neurons. The authors state that anatomical and physiological data suggest that fastigial efferents exert excitatory effects upon spinal motor neurons.

Ascending fastigial fibres project to the superior colliculus, nucleus of posterior commissure, contralateral oral ventral posterior lateral (VPLo) and ventral lateral (VLo) thalamic nuclei (Carpenter and Batton 1982). The greatest number of fibres appear to end in the VPLo. The authors suggest that ascending impulses from the fastigial nucleus are probably relayed to both the primary somaesthetic and motor cortex, where they may influence neurons in the postcentral region representing the neck, limbs and trunk.

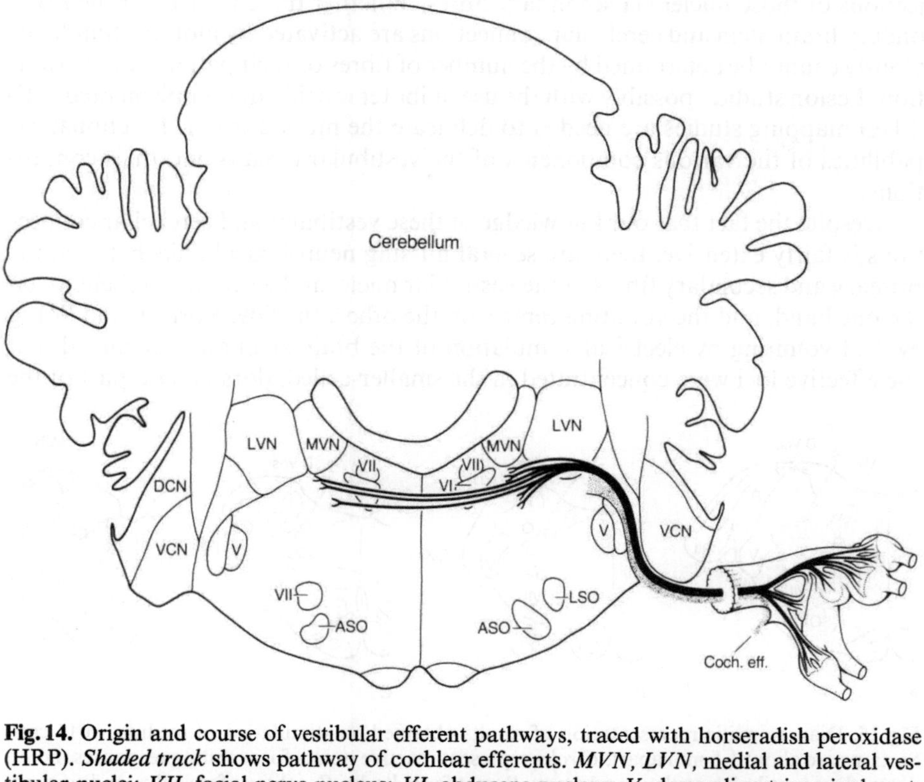

Fig. 14. Origin and course of vestibular efferent pathways, traced with horseradish peroxidase (HRP). *Shaded track* shows pathway of cochlear efferents. *MVN, LVN,* medial and lateral vestibular nuclei; *VII,* facial nerve nucleus; *VI,* abducens nucleus; *V,* descending trigeminal root; *ASO,* accessory superior olivary nucleus; *LSO,* lateral superior olivary nucleus; *VCN,* ventral cochlear nucleus; *DCN,* dorsal cochlear nucleus (From Gacek and Lyon 1974, p 99)

Efferent Vestibular Fibres

According to Gacek (1960) and Gacek and Lyon (1974), the efferent fibres travelling to the labyrinths arise bilaterally from small neurons located lateral to the abducens nucleus and ventral to the MVN (Fig. 14). The fibres merge with the olivocochlear efferent fibres before joining the vestibular root. The efferent fibres become scattered among the afferent fibres and upon penetrating the basement membrane of the sense organs they branch abundantly and provide contacts in the form of vesiculated bouton-type terminals at the base of type II hair cells and on the afferent calyces of type I hair cells in all vestibular sense organs (Brodal and Pompeiano 1957; Gacek 1971).

Vomiting Centre and Parvicellular Reticular Formation

The neuroanatomical studies outlined above, together with those of earlier workers (Brodal and Pompeiano 1957; Brodal and Torvik 1957; Brodal et al. 1962; Brodal and Hoivik 1964; Brodal 1972), present a fairly complete picture of the distribution of primary vestibular fibres in the vestibular nuclei and the projections of those nuclei via secondary fibres. Whether these nuclei and their secondary brain-stem and cerebellar connections are activated by motion stimuli obviously cannot be determined by the number of fibres or their patterns of distribution. Lesion studies, possibly with the use of ibotenic acid, and supplemented with 2-DG mapping studies are needed to delineate the motion emetic functional capabilities of the various components of the vestibular circuits and their connections.

Despite the fact that our knowledge of these vestibular and cerebellar connections is fairly extensive, there are several missing neurological links between the primary and secondary fibres of the vestibular nuclei and vestibulocerebellum, on the one hand, and the vomiting centre, on the other. In 1949, Borison and Wang evoked vomiting by electrical stimulation of the brain stem and concluded that the effective loci were concentrated in the smaller-celled, dorsolateral part of the

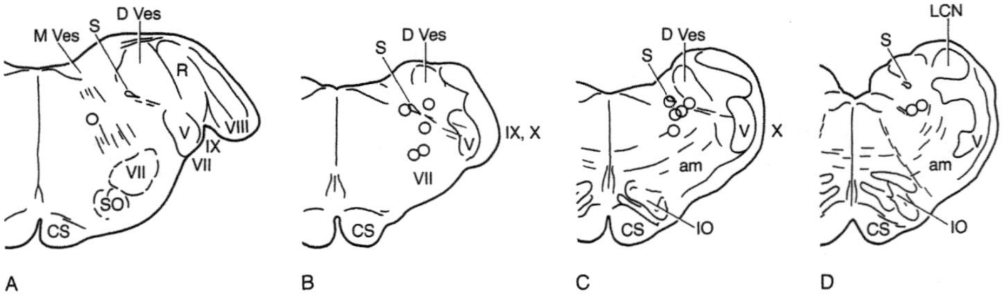

Fig. 15. Diagrammatic representation of results obtained on stimulation of medulla in 11 cats. Sections are about 1.5 mm apart. *Circles* represent reactive points for vomiting responses. *D Ves*, descending vestibular nucleus and tract; *R*, restiform body; *S*, solitary fasciculus and nucleus; *Am*, ambiguus nucleus; *SO*, superior olive; *IO*, inferior olive; *CS*, corticospinal tract; *LCN*, lateral cuneate nucleus. (From Borison and Wang 1949, p 2)

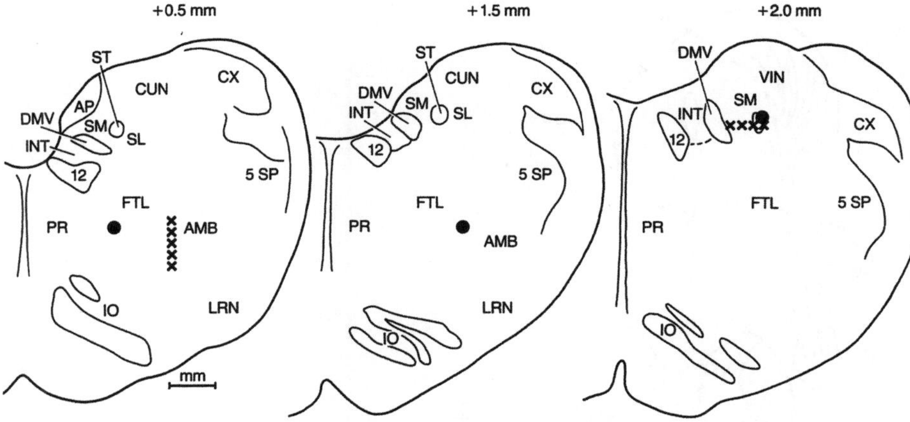

Fig. 16. Sites in which vomiting was produced by electrical stimulation of the brain stem (*filled circles*). Vomiting-like or retching-like behaviour without expulsion of gastric contents was obtained in two other animals at sites marked with *x*. *AMB*, n ambiguus; *AP*, area postrema; *CUN*, cuneate n; *CX*, external cuneate n; *DMV*, dorsal motor n of vagus; *FTL*, lateral tegmental field; *INT*, n intercalatus; *IO*, inferior olive; *LRN*, lateral reticular n; *PR*, paramedian reticular n; *SL*, lateral solitary n; *SM*, medial solitary n; *ST*, solitary tract; *VIN*, descending vestibular n; 5 *SP*, spinal trigeminal n; 12, hypoglossal n. (From Miller and Wilson 1983, p 156)

brain-stem reticular formation (Fig. 15). In subsequent studies these investigators found that destruction of this zone with the implantation of radon seeds resulted in loss of the emetic response to both central and peripheral emetic stimuli in animals previously emetic-sensitive to these stimuli. These data strengthened the concept of the existence of a vomiting centre in this region of the reticular formation.

In more recent studies employing techniques similar to those of Borison and Wang (1949), Miller and Wilson (1983) were able to confirm the finding that vomiting can be elicited by electrical stimulation of the brain-stem reticular formation in unanaesthetized decerebrate cats (Fig. 16). However, the incidence of vomiting was lower and the latency longer in the latter studies. As a result, Miller and Wilson (1983) concluded that a discretely localized vomiting centre does not exist. The fact remains, however, that they did induce vomiting or retching movements or prodromal signs of vomiting with their electrical stimulation procedures in essentially the same region of the reticular formation in which Borison and Wang (1949) produced the emetic reflex in earlier experiments.

This zone, now known as the parvicellular reticular formation (PCRF) lies ventral to the vestibular nuclei, medial to and co-extensive with the elongated nucleus of the spinal tract of the trigeminus and dorsolateral to the larger-celled medullary reticular cell groups which give rise to various reticulospinal pathways. A schematic drawing of the PCRF is shown in Fig. 17. PCRF forms a column that extends from the level of the obex rostrally to the pontine parabrachial region. The motor nucleus of the trigeminus is partially embedded in the oral pole of the PCRF column, and it receives massive ascending axon connections originating from both ipsilateral and contralateral PCRF cells, especially a subset of PCRF

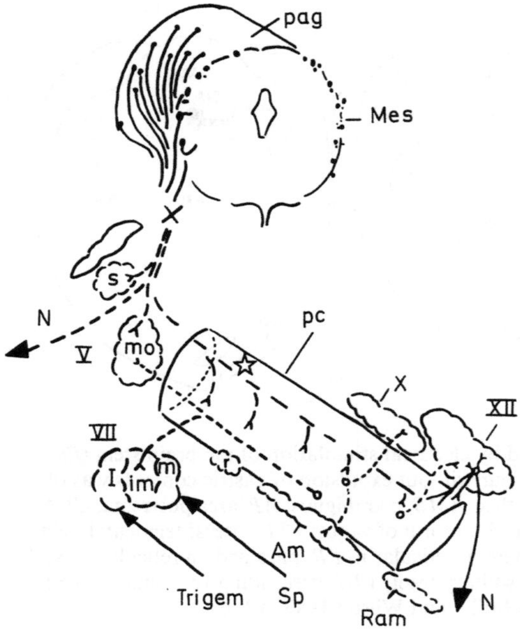

Fig. 17. Schematic perspective drawing of the cylindroid-shaped parvicellular reticular formation (*PC*) showing the relative positions of the cranial nerve nuclei interconnected by cells in the PC, and the distribution of axons and axon collaterals of the mesencephalic root of the trigeminus to the supratrigeminal nucleus (*S*), the motor nucleus of the trigeminus (*Mo*) and the descending bulbar prolongation (Probst's tract) into *PC; star,* relative position of vestibular efferent cell cluster. (From Mehler 1983, p 65)

cells which lie just ventrolateral to the hypoglossal nuclei column in the lower medulla.

Other regions of PCRF also contain cells that send ascending fibre projections to certain subdivisions of the facial motor nucleus. The hypoglossal nucleus also receives PCRF cell connections as described in greater detail in a recent paper by Mehler (1983). These observations suggest the functional importance of subsets of cells in various parts of PCRF as co-ordinators of somatomotor nuclei controlling the muscles of mastication and the tongue muscles. King (1980) showed that axons of a large number of PCRF cells as well as solitary nucleus cells ascend to the parabrachial nuclei at pontine levels that contain certain respiratory "centres". King's analysis of the PCRF projections led him to conclude that the region is organized in discrete longitudinal sheets that run parallel to the axis of the brain stem.

Mehler's (1980) demonstrations of PCRF projections to the solitary and vagal nuclei originating chiefly in a continuous basal layer of the PCRF lends credence to King's notion of a laminar organization (Fig. 18).

According to Holstege and co-worker's (1983) analysis of a swallowing "centre", PCRF also may participate in the control of upper laryngeal muscles, such as the tensor veli palatini muscle, innervated by the trigeminal motor nucleus, which stiffens the soft palate in the initial phase of swallowing and, we

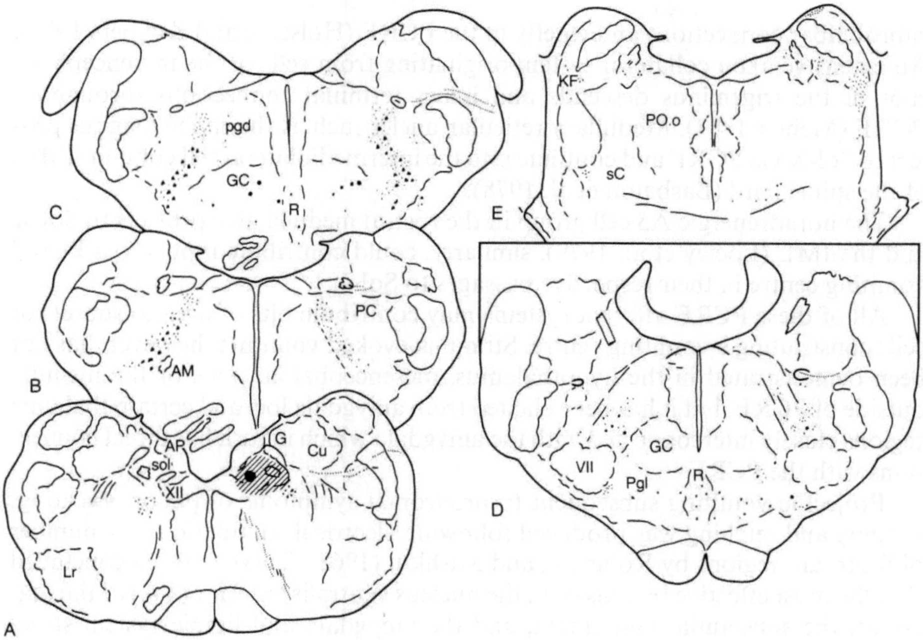

Fig. 18. Labelled small- (*dots*) and medium- (*solid triangles*) sized cell locations after a pipette injection of HRP into the caudal solitary, vagal and hypoglossal nuclei on the right side (*diagonal lines*) in the cat. (From Mehler 1983, p 71)

would surmise, probably also contracts in the vomiting act sequence. Laryngeal and pharyngeal muscles appear to go flaccid in the act of emesis (Smith and Brizzee 1961). Cells of the nucleus ambiguous (Am) whose axons innervate these muscles via the glossopharyngeal and vagus nerves receive their major afferent input from the solitary tract nucleus (Sol), as Beckstead et al. (1980) have re-verified with modern tritiated amino acid axon transport studies. There is no data supporting PCRF cell connections with Am, however.

Afferent Connections of PCRF

There are extensive corticobulbar connections with PCRF originating from motor and sensory cortices (Kuypers 1958). The new anterograde and retrograde axon transport method studies suggest that more viscerally related basal forebrain cortices may also contribute to the corticobulbar projection system that distributes to PCRF in the dorsolateral tegmentum of the brain stem. The new methods, similarly, have revealed that there are certain subcortical forebrain nuclei that project to PCRF. The central nucleus of the amygdala, for example, which projects to the solitary and vagal nuclei (Sol-X) also appears to distribute diffuse endings throughout PCRF in its descent through the medulla (Hopkins and Holstege 1978; Price and Amaral 1981; Schwaber et al. 1982). There are also descending hypothalamic projections to PCRF that terminate chiefly on salivatory nuclei dispersed in the rostral PCRF region (Hosoya et al. 1983). Crossed rubrobulbar connections and chiefly in the PCRF (Holstege and Kuypers 1977). An extensive axon collateral system originating from cells of the mesencephalic root of the trigeminus descends and issues terminal connections throughout PCRF (Mehler 1983). Medullary reticular nuclei such as the raphé magnus project to Sol-X via PCRF and continue into the intermediolateral cell column (IML) of the spinal cord (Basbaum et al. 1978).

The noradrenergic A5 cell group in the rostral medulla also projects to Sol-X and the IML (Loewy et al. 1979), similarly, could contribute inputs to a PCRF vomiting centre in their respective passages to Sol-X.

All of these PCRF afferent systems may contribute bits of data to subsets of cells constituting a vomiting centre. Stimulus-evoked vomiting, however, has not been demonstrated in the hypothalamus, mesencephalon, pons or the medulla outside of PCRF, but it has been elicited from amygdala loci and certain thalamic regions closely interconnected with the amygdala which in turn has direct connections with the PCRF.

Projectile vomiting subsequent to prodromal symptoms of pallor, sweating, gagging and retching was produced following electrical stimulation of a number of forebrain regions by Robinson and Mishkin (1968). These authors concluded that the most effective regions were the nucleus ventralis anterior (VA) of the thalamus, the substantia innominata, and the amygdala and limbic system structures, such as the hippocampus and fornix and the septal region.

We can offer no explanation for the implication of VA as a vomiting response locus based upon the known connections of VA. Anatomically VA has been shown to receive input only from the medial segment of the globus pallidus and substantia nigra and to send its axons to the precentral cortex (Nauta and Mehler 1966). The rationale for the positive results that these authors obtained from stimulation of other basal ganglia structures, such as the putamen, is equally enigmatic. The fundus striati portion of the putamen in which the vomiting response was demonstrated and VA, however, are both traversed by midline thalamic nuclei projections to and/or from the orbitofrontal cortex, basal frontal regions (such as the substantia innominata) and the amygdala (Kievet and Kuypers 1977; Aggleton et al. 1980; Mehler 1980). With the exception of VA, it should be noted that the midline thalamic nuclei were the *only* other diencephalic loci described at the time from which stimulus-evoked vomiting was produced. It is of interest to note further that recent studies of midline thalamic nuclei afferent connections by Aggleton and Mishkin (1984) have shown that these nuclei and the adjacent magnocellular part of the medial dorsal nucleus (MDmc) respectively receive chiefly medial and basal amygdala nuclei projections. These nuclei are the amygdala regions that Robinson and Mishkin (1968) suggested contained the greatest number of vomiting loci.

Anatomical evidence now exists that some cells in the substantia innominata that contains vomiting loci also have direct connections with PCRF and the Sol-X complex (Schwaber et al. 1982). Review of the potential pathways of forebrain limbic system structures reported to contain vomiting loci is beyond the scope of this treatise which is aimed chiefly at localization of vomiting centre in the PCRF of the brain stem. Although the supradiaphragmatic section of the vagus nerve does not prevent motion-induced vomiting (Brizzee 1956), stomach awareness transmitted by the vagus nerve undoubtedly plays some role in normal-gravity motion sickness and also in zero-gravity emesis now called the "space adaptation syndrome", in which case normal-gravity motion sickness prodromal symptoms of sweating, pallor and even nausea may be absent. The literature bearing on the vagaly transmitted increase in gastric motility and acid production and gastric ulcer pathology, induced by electrical stimulation of the centromedial amygdala, is reviewed by Henke (1982). The point is there is now also clear anatomical evidence of direct amygdala projections to the Sol-X complex that also contribute connections to PCRF containing the vomiting "centre". There is a distinct sheet of PCRF cells that was previously mentioned whose axon terminals also converge on the Sol-X complex that one could speculate may inhibit vagal activity and produce the flaccid gastrointestinal tract observed in emesis (Smith and Brizzee 1961).

The last, but not least, (connectional) question to be considered is the possibility of vestibular input into PCRF relayed directly from the vestibular nuclei to a vomiting centre region in PCRF. There is an extensive commissural system interconnecting the vestibular nuclei complexes whose axons traverse a large extent of the underlying PCRF on both sides of the brain stem. Rat Golgi sections and anterograde HRP filling of vestibulofugal axons in one rat brain experiment suggested that there might be very fine axon collaterals issuing from these commissural fibres, or other vestibulofugal axons, that arborized in PCRF (Mehler 1983). Tritiated amino acid studies of vestibular nuclei projections in the cat and monkey, however, have *not* confirmed these suggested PCRF afferent connections (Carleton and Carpenter 1983). There are, therefore, multiple sources of input descending to PCRF from cortical and subcortical regions that present potential somatic and visceral afferent pathways that could transmit information to the hypothesized vomiting integration centre in the PCRF. Which of these pathways are involved and how PCRF orchestrates these pieces of information to control the final common effector pathways is still unknown. Established *efferent* connections made by PCRF cells, as we have noted, are chiefly projections to brain-stem motor nuclei.

Solitary-Vagal Nuclear Complex

Several studies in the past few years have suggested a viscerotopic organization of primary afferent fibres terminating in the solitary nucleus (Sol). Gwyn et al. (1979) employing anterograde and retrograde fibre tracing techniques demonstrated afferent projections of the vagus nerve in the cat to the nucleus of the solitary tract (NTS), including the part designated as the area subpostrema (ASP). A projection to the area postrema (AP) was also demonstrated. Stomach wall injections of horseradish peroxidase (HRP) labelled fibres and terminals bilaterally in the ASP and more ventral parts of the Sol. In 1982, Gwyn et al. in ultrastructural studies following injection of HRP into the nodose ganglion in the cat revealed axon terminals containing clear round vesicles which synapsed upon dendritic elements in the NTS. Leslie et al. (1982a) found that removal of the nodose ganglion in the cat revealed degenerating terminals of vagal afferent fibres throughout the neuropil of the subnucleus gelatinosus of the NTS. Further studies by Leslie et al. (1982b) in the rat revealed that cervical vagus nerve injection of HRP heavily labelled the ipsilateral Sol and commissural nucleus. Lighter labelling was seen contralaterally and this did not extend as far rostrally in the Sol. Labelling was also demonstrated in the area postrema. Retrograde labelling of the neuron somata was observed in the nodose ganglion, throughout the whole extent of the dorsal motor nucleus of the vagus (DXV), much of the nucleus ambiguus (AM) and even in rostral levels of the spinal cord. After stomach injections, labelling was seen in the dorsomedial and medial portions of the Sol and in the commissural nucleus. Labelled efferent fibres were demonstrated from neurons in the DXV, AM and cervical spinal cord.

There is a consensus, however, that vagal cells that project to the stomach are distributed randomly throughout the vagal nucleus column in the rat, cat and monkey (Karim et al. 1984).

From the studies cited above and also the studies of Kalia and Mesulam (1980), it seems clear that gastrointestinal afferent fibres of the subdiaphragmatic vagus nerve terminate chiefly, but not exclusively, in the medial subnucleus of the solitary nucleus (Sol) and in the commissural part of Sol. Kalia and Mesulam (1980) concluded that while prime "cardiovascular" and "respiratory" terminal Sol subnuclei did not receive abdominal visceral input, gastric afferent fibre connections also terminate on the subnucleus gelatinosus in which cardiovascular and respiratory fibres also terminate. Kalia and Mesulam's study (1980) also revealed evidence of some direct sensory fibre projections to the area postrema (Borison and Brizzee 1951) from the bronchi, lungs, heart and stomach.

Immunohistochemical analyses of Sol are also demonstrating new aspects of intra-nuclear organization in Sol in the rat (Kalia et al. 1984) and in the cat (Maley and Elde 1982). For example, the chiefly "gastrointestinal" medial Sol has been shown to contain the most dense substance P terminals and the vagal nucleus is also "rich" in substance P. According to Kalia and co-workers (1984), who call attention to these facts, the role of substance P in modulating gastrointestinal motility should be investigated.

Comments

The recent observations of Miller and Wilson (1983) appear to raise a question as to the existence of a focal vomiting centre in the brain stem. However, it should be noted that Borison and Wang (1949), unlike the former authors, used "stimulus-bound" expulsion of vomitus as the criterion against which the vomiting centre (VC) was localized, and the entire area of stimulation was confined to approximately the lower two-thirds of the zone now recognized as the parvicellular reticular formation.

In a complimentary ablation experiment, it was observed that destruction of this region rendered all experimental animals refractory to all emetic stimuli. These data, together with the recent firm documentation of nerve fibre connections between the PCRF and several brain-stem nuclei and areas constituting integral functional components of various visceral pathways and actual emetic circuits, suggest that the PCRF, whether or not it is considered a vomiting centre per se, is of primary importance as the structural substrate of the final co-ordinating mechanisms involved in the elicitation of the vomiting reflex to all types of emetic stimuli.

This study was supported by NASA/Ames Research grant NAG-2-101 and NIH grant RR 00164-23

References

Aggleton JP, Mishkin M (1984) Projection of the amygdala to the thalamus in the cynomolgus monkey. J Comp Neurol 222:56–68

Aggleton JP, Burton MJ, Passingham RE (1980) Cortical and subcortical afferents to the amygdala of the rhesus monkey (*Macaca mulatta*). Brain Res 190:347–368

Babkin BP, Bornstein MB (1943) The effect of swinging and of binaural galvanic stimulation on the motility of the stomach in dogs. Rev Can Biol 2:336–342

Basbaum AI, Clanton CH, Field HL (1978) Three bulbospinal pathways from the rostral medulla of the cat: an autoradiographic study of pain modulating systems. J Comp Neurol 178:209–224

Beckstead RM, Morse JR, Norgren R (1980) The nucleus of the solitary tract in the monkey. Projections to the thalamus and brainstem nucleus. J Comp Neurol 190:259–282

Bloom W, Fawcett DW (1975) A textbook of histology. Saunders, Philadelphia, p 973

Borison HL, Brizzee KR (1951) Morphology of the emetic chemoreceptor trigger zone in the cat medulla oblongata. Proc Soc Exp Biol Med 77:38–42

Borison HL, Wang SC (1949) Functional localization of central co-ordinating mechanism for emesis in cat. J Neurophysiol 12:305–313

Brizzee KR (1956) Effect of localized brainstem lesions and supradiaphragmatic vagotomy on x-irradiation emesis in monkey. Am J Physiol 187:567–570

Brizzee KR, Dunlap WP (1983) Motion-induced alterations in 2-deoxyglucose uptake in brainstem nuclei of squirrel monkeys: autoradiographic and liquid scintillation studies. Brain Behav Evol 23:14–25

Brodal A (1972) Vestibulocerebellar input in the cat anatomy. In: Brodal A, Pompeiano O (eds) Progress in brain research. Basic aspects of central vestibular mechanisms. Elsevier, Amsterdam, pp 315–321

Brodal A, Hoivik B (1964) Site and mode of termination of primary vestibulocerebellar fibres in the cat. An experimental study with silver impregnation methods. Arch Ital Biol 102:1–21

Brodal A, Pompeiano O (1957) The vestibular nuclei in the cat. J Anat 91:438–454

Brodal A, Torvik A (1957) Über den Ursprung der sekundären vestibulocerebellären Fasern bei der Katze. Eine experimental-anatomische Studie. Arch Psychiatr Nervenkr 195:550–567

Brodal AP, Pompeiani O, Walberg F (1962) The vestibular nuclei and their connections: anatomy and functional correlations. Oliver and Boyd, Edinburgh

Brown GL, McArdle B, Magladery JW (1941) Interim report on clinical investigations into air sickness. Flying Personnel Research Committee Report 410 (a), 19411

Carleton SC, Carpenter MB (1983) Afferent and efferent connections of the medial, inferior and lateral vestibular nuclei in the cat and monkey. Brain Res 278:29–51

Carleton SC, Carpenter MB (1984) Distribution of primary vestibular fibres in the brainstem and cerebellum of the monkey. Brain Res 294:281–298

Carpenter MB, Batton RR (1982) Connections of the fastigial nucleus in the cat and monkey. Exp Brain Res [Suppl] 6:250–295

Copenhaver WM, Bunge RP, Bunge MB (1971) Bailey's textbook of histology. Williams and Wilkins, Baltimore, p 700

Gacek RR (1960) Efferent component of the vestibular nerve. In: Rasmussen GL, Windle WF (eds) Neural mechanisms of the auditory and vestibular system. Thomas, Springfield

Gacek RR (1969) The course and central termination of first order neurons supplying vestibular end organs in the cat. Acta Otolaryngol [Suppl] (Stockh) 254:5–66

Gacek RR (1971) Anatomical demonstration of the vestibulo-ocular projections in the cat. Acta Otolaryngol [Suppl] (Stockh) 293:1
Gacek RR (1978) Location of commissural neurons on the vestibular nuclei of the cat. Exp Neurol 59:479
Gacek RR, Lyon MJ (1974) Localisation of vestibular efferent neurons in the kitten with horseradish peroxidase. Acta Otolaryngol 77:92–101
Graybiel A (1965) Functional disturbance of vestibular origin of significance in space flight. 2nd International symposium on basic environmental problems of man in space. 14–18 June 1965, Paris, France. US Naval Aviation Medical Centre, Pensacola, Fla, pp 1–30
Graybiel A, Miller FF (1970) The otolith organs as a primary aetiological factor in motion sickness with a note on off-vertical rotation. Fourth symposium on the role of the vestibular organs in space exploration. NASA SP-187, Washington, DC, US Government Printing Office, pp 53–64
Gwyn DG, Leslie RA (1979) A projection of vagus nerve to the area subpostrema in the cat. Brain Res 161:335–341
Gwyn DG, Leslie RA, Hopkins DA (1979) Gastric afferents to the nucleus of the solitary tract in the cat. Neurosci Lett 14:13–17
Gwyn DG, Wilkinson PH, Leslie RA (1982) The ultrastructural identification of vagal terminals in the solitary nucleus of the cat after anterograde labelling with horseradish peroxidase. Neurosci Lett 28:139–143
Henke PG (1982) The telencephalic limbic system and experimental gastric pathology: a review. Neurosci Biobehav Rev 6:381–390
Highstein SM (1971) Organization of the inhibitory and excitatory vestibuloocular reflex pathways to the third and fourth nuclei in the rabbit. Brain Res 32:218
Holstege G, Kuypers HGJM (1977) Propriobullar fibre connections to the trigeminal, facial and hypoglossal motor nuclei. I. An anterograde degeneration study in the cat. Brain 100:239–264
Holstege G, Kuypers HGJM, Decker JJ (1977) The organization of the bulbar fibre connections to the trigeminal facial and hypoglossal motor nuclei. II. An autoradiographic tracing study in the cat. Brain 100:265–286
Holstege G, Graveland G, Bijker-Biemond C, Schuddeboom I (1983) Location of motoneurons innervating soft palate, pharynx and upper oesophagus. Anatomical evidence for a possible swallowing centre in the pontine reticular formation. An HRP and autoradiographical tracing study. Brain Behav Evol 23:47–62
Hopkins DA, Holstege G (1978) Amygdaloid projections to the mesencephalon, pons and medulla oblongata in the cat. Exp Brain Res 32:529–549
Hosoya Y, Matsushita M, Sugiura Y (1983) A direct hypothalamic projection to the superior salivatory nucleus neurons in the rat. A study using anterograde autoradiographic and retrograde HRP methods. Brain Res 266:329–333
Johnson WH, Taylor NBG (1961) Some experiments on the relative effectiveness of various types of accelerations on motion sickness. Aerospace Med 32:205–208
Kalia M, Mesulam MM (1980) Brain stem projections of sensory and motor components of the vagus complex in the cat. II. Laryngeal, tracheobronchial, pulmonary, cardiac and gastrointestinal branches. J Comp Neurol 193:467–508
Kalia M, Fuxe K, Hokfelt T, Johansson O, Land R, Ganten D, Cuello C, Terenius L (1984) Distribution of neuropeptide immunoreactive nerve terminals within the subnuclei of the nucleus of the tractus solitarius of the rat. J Comp Neurol 222:409–444
Karim MA, Shaikh E, Tan J, Ismail Z (1984) The organization of the gastric efferent projections in the brain stem of monkey: HRP studies. Brain Res 293:231–240
Kennedy C, Sakuroda O, Shinohara M, Jehle JW, Sokoloff L (1978) Local cerebral glucose utilization in the normal conscious macaque monkey. Ann Neurol 4:293–301
Kievet J, Kuypers HGJ (1977) Organization of thalamocortical connections to frontal lobe in rhesus monkey. Exp Brain Res 29:299–322
King GW (1980) Topology of ascending brain stem projections to nucleus parabrachialis in the cat. J Comp Neurol 191:615–638
Kurashvili AE (1964) Vestibular reactivity during the cumulative action of slow centripetal accelerations. Zh Ushnykh Nosovykh I Gorlovnykh Bolezney 1:49–55 (in Russian)
Kuypers HGJM (1958) Some projections from the pericentral cortex to the pons and lower brainstem in monkey and chimpanzee. J Comp Neurol 110:221–255

Leslie RA, Gwyn DG, Morrison CM (1978) The fine structure of the ventricular surface of the area postrema in the cat, with particular reference to supraependymal structures. Am J Anat 153:273–290
Leslie RA, Gwyn DG, Hopkins DA (1982a) The ultrastructure of the subnucleus gelatinosus of the nucleus of the tractus solitarius in the cat. J Comp Neurol 206:109–118
Leslie RA, Gwyn DG, Hopkins DA (1982b) The central distribution of the cervical vagus nerve and gastric afferent and efferent projections in the rat. Brain Res Bull 8:37–43
Lindeman HH (1973) Anatomy of the otolithorgans. Adv Otorhinolaryngol 20:405–433
Loewy AD, McKellar S, Saper CB (1979) Direct projections from the A5 catecholamine cell group to the intermediolateral cell column. Brain Res 174:309–314
McNally WJ, Stuart EA (1942) Physiology of the labyrinth in relation to sea sickness and other forms of motion sickness. War Med 2:683
Maley B, Elde R (1982) Immunohistochemical localization of putative neurotransmitters within the feline nucleus tractus solitarii. Neuroscience 7:2469–2490
Mehler WR (1980) Subcortical afferent connections of the amygdala in the monkey. J Comp Neurol 190:733–762
Mehler WR (1983) Observations on the connectivity of the parvicellular reticular formation with respect to a vomiting centre. Brain Behav Evol 23:63–80
Miller AD, Wilson VJ (1983) Vomiting Centre reanalyzed: an electrical stimulation study. Brain Res 270:154–158
Miller EF, Graybiel A (1972) Semicircular canals as a primary etiological factor in motion sickness. Aerospace Med 43:1065–1074
Money KE, Friedburg J (1964) The role of the semi-circular canals in causation of motion sickness and nystagmus in the dog. Can J Physiol Pharmacol 42:793–801
Nauta WJH, Mehler WR (1966) Projections of the lentiform nucleus in the monkey. Brain Res 1:3–42
Price JL, Amaral DG (1981) An autoradiographic study of the projections of the central nucleus of the monkey amygdala. J Neurosci 1:1242–1259
Robinson BW, Mishkin M (1968) Alimentary responses to forebrain stimulation in monkeys. Exp Brain Res 4:330–366
Ryu JH, McCabe BF (1973) Neural activity in the vestibular nuclei of the cat. Ann Otol Rhinol Laryngol 82[Suppl 9]:3–28
Schwaber JS, Kapp BS, Higgins GA, Rapp PR (1982) Amygdaloid and basal forebrain direct connections with the nucleus of the solitary tract and the dorsal motor nucleus. J Neurosci 2:1424–1438
Sjoberg AA (1931) Experimentelle Studien über den Auslösungsmechanismus der Seekrankheit. Acta Otolaryngol [Suppl] (Stockh) 14:1–136
Smith CC, Brizzee KR (1961) Cineradiographic analysis of vomiting in the cat. I. Lower oesophagus, stomach and small intestine. Gastroenterology 40:654–664
Sokoloff L, Reivich M, Kennedy C, Des Rosiers MH, Patlak CS, Pettigrew KD, Sakurads O, Shinohara M (1977) The [^{14}C] deoxyglucose method for the measurement of local cerebral glucose utilization: theory, procedure, and normal values in the conscious and anesthetized albino rat. J Neurochem 28:897–916
Spoendlin H (1966) Some morphofunctional and pathological aspects of the vestibular sensory epithelia. Second symposium on the role of the vestibular organs in space exploration. NASA 5P-115, Washington, DC, US Government Printing Office, pp 115
Tarlov E (1970) Organization of vestibulo-oculomotor projections in the cat. Brain Res 20:159
Wang SC, Chinn HI (1956) Experimental motion sickness in dogs: importance of labyrinth and vestibular cerebellum. Am J Physiol 185:617–623
Wersall J (1960) Electron micrographic studies of vestibular hair cell innervation. In: Rasmussen GL, Windle WF (eds) Neural mechanisms of the auditory and vestibular systems. Thomas, Springfield, Ill, pp 247–257
Wersall J (1966) Studies on the structure and innervation of the sensory epithelium of the cristae ampullares in the guinea pig. A light and electron microscopic investigation. Acta Otolaryngol [Suppl] (Stockh) 126:1–85
Williams PL, Warwick R (1975) Functional anatomy of man. Saunders, Philadelphia
Wruble LD, Rosenthal RH, Webb WL Jr (1982) Psychogenic vomiting. A review. Am J Gastroenterol 77:318–321

Peripheral Nervous Pathways Involved in Nausea and Vomiting

J. L. WILLEMS [1] and R. A. LEFEBVRE [1]

Introduction

During nausea and vomiting, a complex sequence of autonomic and somatic phenomena is observed: changes in respiratory rhythm and depth, increase in salivation, licking and deglutition, gastric relaxation and duodenal-pyloric spasm, contractions of the respiratory muscles and changes in heart rate and blood pressure. The activation of the efferent neurons initiating these events is co-ordinated by the vomiting centre located in the medulla oblongata, in the floor of the fourth ventricle. The vomiting centre receives input from the cerebral cortex, the vestibular organs, the chemoreceptor trigger zone (CTZ) and from afferent fibres coming from the periphery, including the gastrointestinal system. We will discuss the peripheral afferent and efferent nervous pathways involved in nausea and vomiting, focusing our attention on the nervous pathways to and from the gastrointestinal system. As it has been suggested that dopamine plays a role in the pathways controlling the gastric relaxation which occurs during nausea, this will be discussed in more detail. For further information and references about the subject discussed, the reader is referred to the general literature list given at the end of this chapter. Within the text references are cited only for some particular experimental observations.

Afferent Nervous Pathways

Visual, olfactory and gustatory stimuli, and sensory stimuli from almost any site in the body can elicit nausea and vomiting. They are conveyed to the central nervous system (CNS) via the afferent pathways conducting sensory information.

The gastrointestinal system is an important source of sensory stimuli. In the gastrointestinal system several neural plexuses exist which are essentially similar in pattern throughout the tubular digestive system in all species examined. A subserous plexus containing only a few ganglia is found under the serosa; between the longitudinal and circular smooth muscle layer lies the myenteric or Auerbach's plexus containing prominent ganglia and nerve strands; a muscular plexus containing a few ganglia is situated deep in the circular muscle layer; finally the submucous or Meissner's plexus, sometimes consisting of two interconnected plexuses, is found under the mucosa. The intrinsic plexuses with their afferent and efferent pathways are illustrated in Fig. 1.

1 J. F. and C. Heymans Institute of Pharmacology, University of Ghent Medical School, De Pintelaan 185, B-9000 Ghent, Belgium

Nausea and Vomiting: Mechanisms and Treatment
Ed. by C. J. Davis, G. V. Lake-Bakaar and D. G. Grahame-Smith

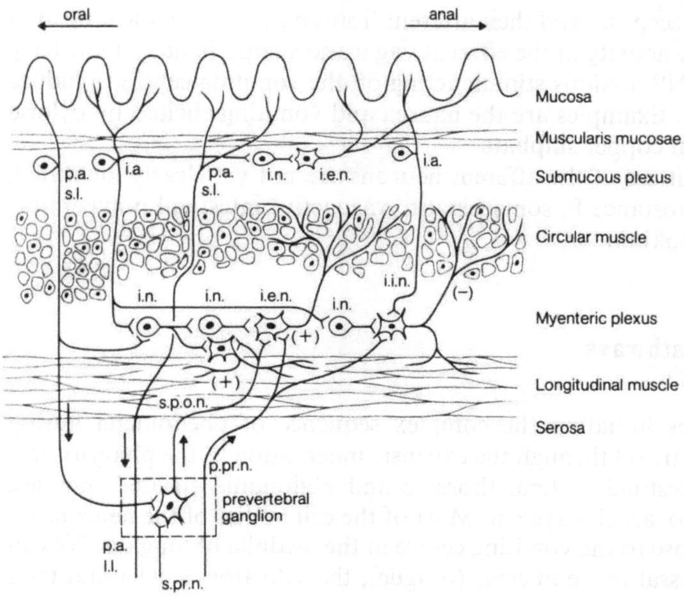

Fig. 1. Schematic portrayal of the various extrinsic and intrinsic enteric neurons. +, excitatory; −, inhibitory; *ia,* intrinsic afferent; *pasl,* primary afferent short loop; *pall,* primary afferent long loop; *in,* interneuron; *ien,* intrinsic excitatory neuron; *iin,* intrinsic inhibitory neuron; *sprn,* sympathetic preganglionic neuron (or nerve); *spon,* sympathetic postganglionic neuron (or nerve); *pprn,* parasympathetic preganglionic neuron (or nerve). [According to Baumgarten (1982), with the permission of the author and the editor]

Intrinsic afferent mechanoreceptive neurons, sensitive to passive distension and active contractions, are found in the submucous plexus. Their axons synapse with interneurons and excitatory and inhibitory neurons situated in the submucous and myenteric plexuses, but also with sympathetic cells in the prevertebral ganglia which belong to the extrinsic neural system. These sensory neurons play a role in the short loop reflexes controlling the inhibitory and excitatory waves of the peristaltic reflex.

Besides the mechanoreceptive neurons in the submucous plexus, there are chemoreceptors, sensitive to food constituents, mechanoreceptors and nociceptors, linked to extrinsic afferent neurons whose cell bodies are situated in the vagal sensory ganglia and in the sympathetic spinal dorsal root ganglia. Sensory oesophageal fibres, mostly showing free endings, run in the vagus, the glossopharyngeal and in the maxillary branch of the trigeminal nerve. The afferent fibres of stretch receptors in the gastric muscular coat, near the lower oesophageal sphincter and in the antrum, and chemoreceptors in the mucosa send their nerve fibres to the vagus.

Mechanosensitive endings in the lower gastric body and antrum send their fibres in the thoracic and lumbar splanchnic nerves to the dorsal root ganglia of T_7–T_{11}. Sensitive neurons lower down in the gastrointestinal tract send fibres along the hypogastric and pelvic sympathetic nerves and the pelvic parasympathetic system. These receptors and their afferent fibres play a role in the long loop reflexes regulating the activity in the efferent vagal and sympathetic system. They also convey to the CNS noxious stimuli acting on the vomiting centre to induce nausea and vomiting. Examples are the nausea and vomiting elicited by pyloric distension and by oral copper sulphate.

The neurotransmitters of the afferent neurons are not yet clearly identified, but candidates are substance P, somatostatin, vasoactive intestinal polypeptide, and gastrin/cholecystokinin.

Efferent Nervous Pathways

The efferent impulses initiating the complex sequence of phenomena during nausea and vomiting travel through the extrinsic innervation to the pharynx, larynx, upper gastrointestinal system, thoracic and abdominal striated muscles, diaphragm and cardiovascular system. Most of the cell bodies of the neurons involved are located close to the vomiting centre in the medulla oblongata. We can mention the hypoglossal nerve nucleus (tongue), the salivatory nuclei and their axons (salivary glands), the nucleus ambiguus with its axons in the glossopharyngeal nerve and vagus (muscles of the pharynx), the dorsal nucleus of the vagus with its axons to the gastrointestinal tract, and the vasomotor centre with parasympathetic and sympathetic efferents. The cell bodies of the neurons of the phrenic nerve and intercostal nerves are located in the spinal medulla, but they receive impulses from the respiratory centre, also located close to the vomiting centre.

The extrinsic efferent neural control of the gastrointestinal system consists firstly of excitatory cholinergic preganglionic fibres in the vagal nerves and in the pelvic nerves. They innervate both cholinergic and non-cholinergic intrinsic excitatory motor neurons in the enteric plexuses throughout the gastrointestinal tract, but primarily in the stomach and in the large intestine.

The neurons innervate the circular and longitudinal smooth muscle layer. The non-cholinergic excitatory transmitter might be substance P or a 5-HT related substance. intramural interneurons sometimes intervene in this vagal pathway. The vagal activity transmitted through these pathways seems to stimulate the spontaneous activity of the smooth muscle cells.

The vagal nerves further contain cholinergic preganglionic fibres which innervate intrinsic inhibitory neurons in the enteric plexus. These inhibitory neurons release an unknown neurotransmitter.

The sympathetic system also contributes to the extrinsic neural control of the gastrointestinal system. The preganglionic fibres synapse in the paravertebral or prevertebral ganglia, e.g. the coeliac ganglion for the stomach and duodenum. The postganglionic adrenergic fibres, running along the blood vessels, synapse in the intramural ganglia in the myenteric and submucous plexus. They mainly maintain a presynaptic inhibitory tone on the release of cholinergic transmitter in the enteric ganglia and thereby decrease smooth muscle activity. Only a few fibres innervate the muscular coats or muscularis mucosae and act directly on the smooth muscle cells producing muscle relaxation via postsynaptic β-adrenoceptors or, in the sphincters, muscle contraction via postsynaptic α-adrenoceptors.

Gastric Relaxation During Nausea

In experimental animals and in man, gastric relaxation is observed in the period preceding the evacuation of the stomach contents, i.e. during nausea. This relaxation is brought about by stimuli descending in that part of the vagus which innervates the intrinsic inhibitory enteric neurons. As it was shown that the transmitter released is neither a catecholamine nor acetylcholine, this entity has been named the "non-adrenergic, non-cholinergic inhibitory vagal system". This system is the efferent part of the long loop intestinal reflex which produces gastric receptive relaxation and inhibition of enteric tone during food intake. Several substances have been proposed as neurotransmitters for these inhibitory enteric neurons, e.g. adenosine 5′-triphosphate (ATP), dopamine and vasoactive intestinal polypeptide (VIP).

ATP, or a related purine compound, has been detected in enteric neurons, its synthesizing machinery is present and it is released from intramural nerves; exogenous ATP mimics the relaxatory effect of the inhibitory neurons in some preparations. On the basis of this hypothesis, the inhibitory system has been called "purinergic". However, ATP is synthesized and stored in a large variety of nerve cells, its electrophysiological and mechanical effects on smooth muscle activity are not always similar to those of inhibitory neuron stimulation and the effects of ATP and inhibitory nerve stimulation are not similarly affected by antagonistic drugs. Therefore, there is serious doubt that ATP or a related nucleotide is the transmitter released from enteric inhibitory nerves.

Dopamine also has been proposed as an inhibitory neurotransmitter in the gastrointestinal tract. This is based on several observations showing that dopamine mimics the hyperpolarizing and relaxatory effects of interneuron stimulation. Dopamine has also been proposed as the neurotransmitter of the intrinsic inhibitory neurons of the vagal inhibitory system, since dopamine produces gastric relaxation in vivo and since this effect is antagonized by dopamine antagonists which also antagonize gastric receptive relaxation induced by gastric distension or feeding. Finally, the possibility that dopamine plays a physiological inhibitory role in the regulation of gastrointestinal motility has been raised by the suggestion of the presence of dopamine receptors in the lower oesophageal sphincter, stomach, pylorus and duodenum.

The major criticism for a physiological inhibitory role for dopamine comes from the observations that, after extrinsic denervation, neither dopamine (or another catecholamine) nor tyrosine hydroxylase activity can be found in the gastrointestinal wall, suggesting that the dopamine detected comes from postganglionic extrinsic sympathetic nerves.

Furthermore, we have shown in dogs that the dopamine agonist apomorphine produces prodromal gastric relaxation by stimulation of a specific dopamine receptor on the CTZ (Blancquaert et al. 1982; Lefebvre et al. 1981); this stimulus reaches the stomach via the vomiting centre and the inhibitory vagus. Indeed, apomorphine produces the whole phenomenon of nausea and vomiting via the CTZ.

As the CTZ lies outside the blood-brain barrier and is easily attainable even for substances which do not readily cross into the CNS, e.g. dopamine and the dopamine antagonist domperidone (Fig. 2), one must be careful before concluding from systemic injections of drugs that the effect is due to a direct action on the gastrointestinal tract.

If dopamine is the transmitter of the intrinsic inhibitory neurons, the dopamine antagonist domperidone should also block the gastric relaxatory effects of other emetic stimuli, which produce gastric relaxation via the inhibitory vagus. We have shown that gastric relaxation by morphine is also produced, at least partially, via the CTZ but that domperidone does not influence the gastric relaxatory effect of morphine (Fig. 2) (Blancquaert et al. 1982; Lefebvre et al. 1981). Furthermore, the gastric relaxation preceding the vomiting elicited by Veratrum alkaloid, which acts via the ganglion nodosum and the vomiting centre, is likewise not antagonized by domperidone (unpublished results).

It should be added that the gastric relaxation and vomiting by apomorphine and morphine can be blocked selectively by dopamine antagonists, on the one hand, and naloxone, on the other hand (Fig. 2). It is, therefore, unlikely that dopamine or dopamine receptors play a fundamental role in the pathway from the CTZ to the vomiting centre or in the efferent pathways that start from the vomiting centre.

Finally, as far as the postulated gastric dopamine receptor is concerned, systematic studies in vitro in isolated strips of canine, rat and guinea-pig stomach showed that, with minor species variations, dopamine produces a direct smooth muscle relaxation via postsynaptic β-adrenoceptors and inhibits cholinergic tone by presynaptic inhibition of transmitter release via α-adrenoceptors (Fig. 3) (Lefebvre et al. 1983, 1984). It was further shown that dopamine antagonists like domperidone are not selective enough to be used in experiments as sole markers for the presence of dopamine receptors. Furthermore, for some dopamine antagonists, e.g. metoclopramide, effects different from the dopamine receptor blocking action are present.

The last candidate as the neurotransmitter for the non-adrenergic, non-cholinergic vagal inhibitory system is vasoactive intestinal polypeptide (VIP). A VIP-like substance has been detected using immunohistochemical methods in a large number of enteric neurons and axons. This substance is released in the venous effluent of the stomach upon stimulation of the vagal nerves and its release is blocked by the ganglion blocking agent hexamethonium. The VIP-like substance is also released when the inhibitory vagal system is activated by distension of the oesophagus. Intra-arterial infusion of VIP into the stomach produces gastric relaxation. Nevertheless, not all observations are compatible with this hypothesis and final proof that VIP is indeed the inhibitory gastrointestinal neurotransmitter is still not available.

It might be possible that besides the non-adrenergic, non-cholinergic vagal inhibitory system, other neurons also play a role in the efferent pathway of the reflex gastric relaxation. It has been suggested that extrinsic 5-HT containing

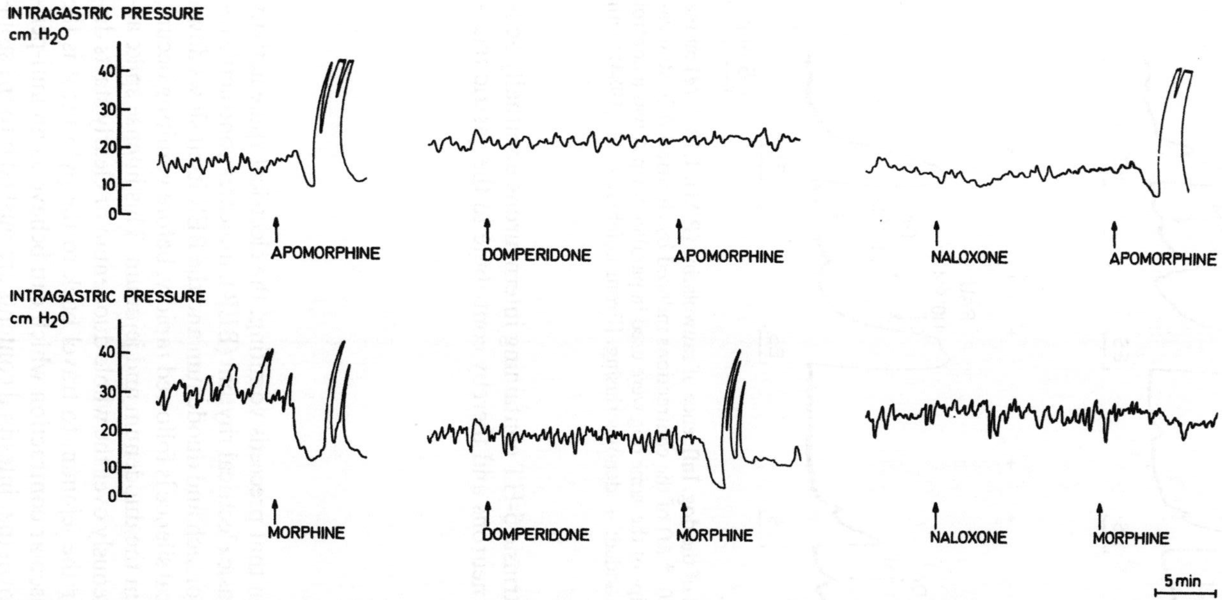

Fig. 2. Influence of subcutaneous apomorphine (0.03 mg/kg) and morphine (0.5 mg/kg) on intragastric pressure in the conscious dog. The second injection of apomorphine and morphine was given 15 min after domperidone (1 mg/kg IV), the third injection 15 min after naloxone (0.07 mg/kg IV). [From Willems et al. (1982), with the permission of the editor]

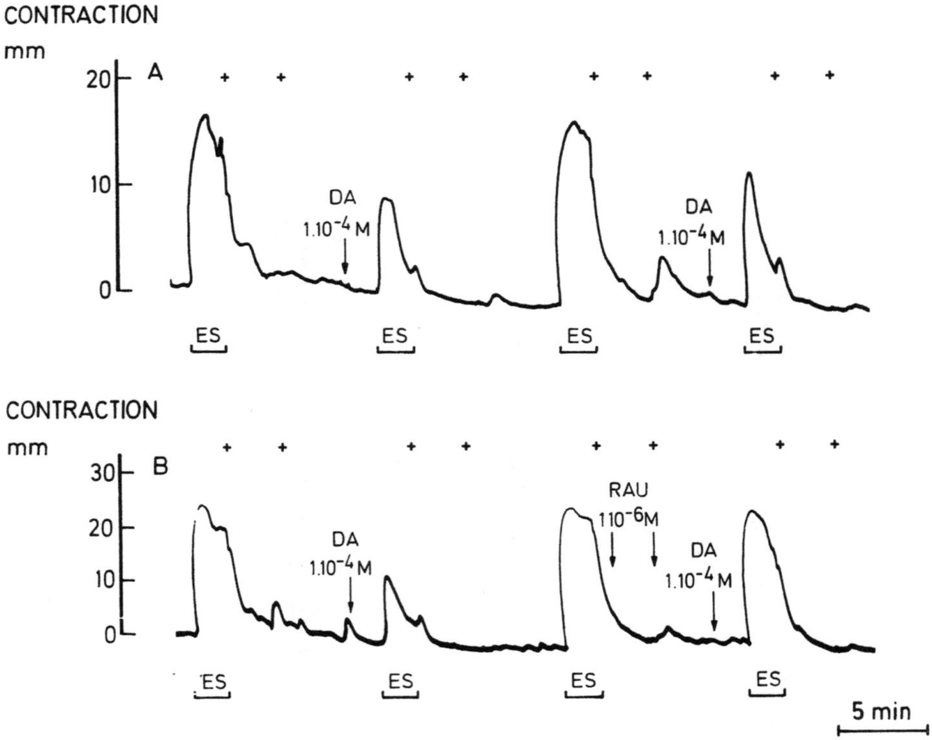

Fig. 3. Longitudinal fundus strips of the dog. Influence of rauwolscine (RAU, 1.10^{-6} *M*) on the inhibition by dopamine (DA, 1.10^{-4} *M*) of the contractions induced by electrical field stimulation (ES, 0.5 Hz, 2 min). Two strips of the same dog were used in parallel. Strip A was a control strip, in strip B rauwolscine was added; + denotes rinsing. [From Lefebvre et al. (1984), with permission of the editor]

neurons in the vagus and intrinsic 5-HT containing interneurons eventually activate the intrinsic inhibitory neurons and thereby contribute to the gastric relaxation.

Anti-peristalsis

During the gastric relaxation that preceeds vomiting, the electrical spike activity that, superimposed on the basic electrical rhythm (BER), indicates contractile behaviour, disappears in the stomach and duodenum and the BER itself slows down or is suppressed. The electrical silence is followed rapidly, before vomiting occurs, by an intense spike activity in the duodenum and jejunum. This intense spike activity either occurs simultaneously over the whole duodenum or clearly starts distally from the duodenum or the jejunum to travel back to the pylorus; it is the marker of a generalized muscular contraction which can behave as an anti-peristaltic wave. Indeed, it is known that intestinal contents are pushed into the stomach during gastric relaxation and are evacuated together with the gastric contents during vomiting.

The normal peristaltic wave in the gastrointestinal tract is regulated by short loop reflexes in which the intrinsic excitatory cholinergic neurons are responsible for the excitatory component and the intrinsic inhibitory neurons for the inhibitory component. It has been suggested that the intense duodenal activity, which always follows gastrointestinal inhibition, is nothing else than the result of the abrupt ending of the inhibitory stimulus. Duodenal spasm or anti-peristalsis does not occur after hexamethonium or vagal section, when gastric relaxation also does not occur. However, anti-peristalsis and duodenal spasm remain present after atropine, which suggests that this might also be due to activation of an intrinsic non-cholinergic control system. Some preliminary experimental findings suggest that a 5-HT-like substance may play a role in it. Further research is needed to explain the anti-peristaltic mechanism.

Summary

Among the afferent pathways impinging on the vomiting centre, some come from the gastrointestinal system and run in the vagal nerves and the sympathetic nerves. Noxious stimuli from the gastrointestinal tract can thereby elicit nausea and vomiting. The vomiting centre activates a multitude of efferent neural pathways innervating the different systems (upper gastrointestinal tract, respiratory muscles and cardiovascular system), which contribute to the complex phenomenon of vomiting.

Particular attention has been paid to the efferent neural fibres and cells that bring about gastric relaxation and duodenal anti-peristalsis that are observed prior to vomiting, during the phase of nausea. The prodromal gastric relaxation is elicited by nerve stimuli travelling down a particular group of preganglionic vagal fibres which innervate in the gastric wall intrinsic inhibitory neurons that produce smooth muscle relaxation. These neurons liberate an unknown neurotransmitter which has been suggested to be ATP, dopamine or VIP.

It is not yet known which extrinsic neural control system is responsible for the anti-peristaltic wave that travels to the stomach just before the gastric evacuation occurs during vomiting.

Further General Reading

Abrahamsson H (1973) Studies on the inhibitory nervous control of gastric motility. Acta Physiol Scand [Suppl] 390

Baumgarten HG (1982) Morphological basis of gastrointestinal motility: structure and innervation of gastrointestinal tract. In: Bertaccini G (ed) Mediators and drugs in gastrointestinal motility. Springer, Berlin Heidelberg New York, pp 7–53 (Handbook of experimental pharmacology, vol 59, part 1)

Borison HL, Wang SC (1953) Physiology and pharmacology of vomiting. Pharmacol Rev 5:193–230

Daniel EE (1982) Pharmacology of adrenergic, cholinergic, and drugs acting on other receptors in gastrointestinal muscle. In: Bertaccini G (ed) Mediators and drugs in gastrointestinal motility. Springer, Berlin Heidelberg New York, pp 249–322 (Handbook of experimental pharmacology, vol 59, part 2)
Furness JB, Costa M (1980) Types of nerves in the enteric nervous system. Neuroscience 5:1–20
Furness JB, Costa M (1982) Identification of gastrointestinal neurotransmitters. In: Bertaccini G (ed) Mediators and drugs in gastrointestinal motility. Springer, Berlin Heidelberg New York, pp 383–462 (Handbook of experimental pharmacology, vol 59, part 1)
Roman C (1982) Nervous control of esophageal and gastric motility. In: Bertaccini G (ed) Mediators and drugs in gastrointestinal motility. Springer, Berlin Heidelberg New York, pp 223–278 (Handbook of experimental pharmacology, vol 59, part 1)
Willems JL, Buylaert WA, Lefebvre RA, Bogaert MG (1985) Neuronal dopamine receptors on autonomic ganglia and sympathetic nerves and dopamine receptors in the gastrointestinal system. Pharmacol Rev (in press)

References

Blancquaert JP, Lefebvre RA, Willems JL (1982) Gastric relaxation by intravenous and intracerebroventricular administration of apomorphine, morphine and fentanyl in the conscious dog. Arch Int Pharmacodyn Ther 256:153–154
Lefebvre RA, Willems JL, Bogaert MG (1981) Gastric relaxation and vomiting by apomorphine, morphine and fentanyl in the conscious dog. Eur J Pharmacol 69:139–145
Lefebvre RA, Blancquaert JP, Willems JL, Bogaert MG (1983) In vitro study of the inhibitory effects of dopamine on the rat gastric fundus. Arch Pharmacol 322:228–236
Lefebvre RA, Willems JL, Bogaert MG (1984) Inhibitory effect of dopamine on canine gastric fundus. Arch Pharmacol 326:22–28
Willems JL, Lefebvre RA, Bogaert MG (1982) Pharmacology of gastric motility. Acta Gastroenterol Belg 45:471–475

The Organisation of Vomiting as a Protective Reflex:

A commentary on the first day's discussions

C. J. DAVIS [1], R. K. HARDING [2], R. A. LESLIE [3], and P. L. R. ANDREWS [4]

Introduction

This symposium has shown that although there is a wealth of information available concerning the nature of vomiting, no consensus yet exists concerning the detailed mechanisms of the initiation and control of most clinically relevant forms of vomiting. There remain fundamental gaps in our knowledge and the symposium has played a valuable role in identifying and highlighting them. Further, it became evident during the course of the discussion periods that there were several themes or concepts in particular upon which it would be worth focussing further study and which could lead to valuable new insights.

This chapter explores those foci which were identified as a result of both the invited presentations and the contributions from the floor. It arose from an editorial decision to give an overview of each day's proceedings and an expressed desire on the part of a number of participants for us to go beyond a straightforward report of the discussion alone. We have tried to set the points in context by providing background data and some additional ideas concerning their possible significance. Those of us who met together to talk for two further days following the meeting in order to attempt this synthesis, hope that this discussion will help to point the way to future experiments and clinical trials which will further our knowledge of the underlying mechanisms of nausea and vomiting.

Key Issues

Some major and as yet unanswered questions which are central to the whole problem of vomiting have been emphasised as a result of the symposium:

1. What is the function of vomiting, particularly in man?
2. What are the relevant features of a vomiting stimulus?
3. What are the sites for the reception of vomiting stimuli and does the area postrema play more than a chemoreceptive role?

1 Department of Clinical Pharmacology, University of Oxford, Radcliffe Infirmary, Woodstock Road, Oxford OX2 6HE, UK
2 Nuclear Effects Section, Protective Sciences Division, Department of National Defence, Defence Research Establishment Ottawa, Shirley Bay, Ottawa, Ontario, Canada KIA 024
3 Department of Anatomy, Faculty of Medicine, Dalhousie University, Halifax, Nova Scotia B3H 4H7, Canada
4 Department of Physiology, St. George's Hospital Medical School, University of London, Cranmer Terrace, Tooting, London SW17 ORE, UK

Nausea and Vomiting: Mechanisms and Treatment
Ed. by C. J. Davis, G. V. Lake-Bakaar and D. G. Grahame-Smith

4. What structural features of the area postrema are important to its' function?
5. How are the various responses involved in nausea and vomiting integrated within the central nervous system?

These questions are addressed below and where possible answers proposed on the basis of the data presented and the concepts discussed during the symposium. Ultimately, solutions to these questions will hopefully facilitate the design of more specific and effective anti-emetic drugs, and help alleviate this most distressing symptom and ubiquitous medical problem.

Why Does Man Vomit?: An Hypothesis Involving the Hierarchical Organisation of a Toxin Defence System

There are several mechanisms by which an animal may protect its body, and most importantly the CNS, from damage by ingested toxins. These may be placed in order of temporal effectiveness, the most immediately effective coming first:

1. The smell or taste of potential foodstuffs which may be avoided due to innate or learned behaviour
2. The detection of toxins by receptors in the gut followed by a central reflex producing an appropriate response; nausea to prevent further consumption, inhibition of gastric motility to confine the toxin to the stomach, and, if necessary, vomiting to purge the system of the already ingested (but not necessarily yet absorbed) toxin
3. The placing of a sensor located within the CNS to detect circulating toxins. This would provide a vomiting signal probably followed by central integration ultimately leading to vomiting.

An hierarchical arrangement (see Table 1) emerges from a consideration of this system which is based on data presented and discussed during the symposium. It may be seen that this goes some way to explain the disparate responses of different species to similar stimuli and perhaps similar responses of individuals to different stimuli. For instance, it is generally accepted that laboratory rats do not vomit; they do, however, develop a profound conditioned taste aversion (CTA) to wide-ranging stimuli (including X-irradiation and toxins) which actually cause vomiting in other species (Coil and Norgren 1981). (CTA is mediated as we know by the AP, the most likely candidate for the central receptor.) The omnivorous rat displays a bias, probably established by natural selection, to associate gustatory and olfactory cues with internal malaise even when these stimuli are separated by long time periods (Garcia et al. 1966; Rozin and Kelat 1971). The first line of defence in the rat (i.e. smell, taste) appears to be extremely highly developed (Roper 1984) and, therefore, perhaps as a result the third and last line has become redundant and is non-functional. In laboratory rats, even the second line of defence seems to be obsolescent because although it is known that gastric afferents in the rat respond to intragastric copper sulphate (Clarke and Davison 1978), vomiting does not result. The detection apparatus is still present, but the vomiting response is absent. Essentially, here we have an example where the primary defence mechanism is so highly tuned that the animal no longer needs to

Table 1. The hierarchical organisation of the defence mechanism for protection of the organism against toxins

Level of defence mechanisms	Location of toxin	Type of sensor and location	Effect	Resultant action
First-level defence	External to gastro-intestinal tract	1. Smell and taste 2. Peripheral	(a) Avoidance (b) Nausea	(a) Learned aversion for potential toxins
Second-level defence	Intra-gastric	1. Gastric chemo-receptors	(a) Nausea (b) Gastric motility ↓	(a) Learned aversion
		2. Near the absorptive site	(c) Avoidance (d) Vomiting	(b) Confining toxin to one area (c) Ejection of toxin
Third-level defence	Within the vascular system	1. Area postrema (CTZ) 2. Within the CNS	(a) Nausea (b) Gastric motility ↓ (c) Vomiting	(a) Ejection of toxin (b) Confining toxin (c) Learned aversion

rely on the other two lines of defence (so that the latter are no longer even effective in this animal).

Let us turn our attention for a moment to another commonly used animal model for vomiting in man, the dog. We know that the rôle of the gastric afferents is still effective in this animal as evidenced by the vomiting resulting from the administration of intragastric copper sulphate after ablation of the area postrema. Following de-afferentation, however, a much higher dose of copper sulphate is needed to elicit vomiting (Wang and Borison 1951), the chemical stimulus then probably acting through a sensor which resides in the CNS. Such species-specific physiological responses illustrate the importance of choosing the correct animal model to be used for the study of vomiting.

Are There Analogous Systems Within the Body that Could be Viewed in the Same Light We Have Employed to Analyse the Vomiting Defence System?

As we have seen, a three-tier hierarchical system may guard against the ingestion of toxins. The respiratory system, too, is involved in taking into the body exogenous substances. It can be argued that this system utilizes a similar three-tier protection mechanism (Table 2).

Table 2. A hierarchical organisation of the respiratory defence mechanism

Defence Level	Sensor and Location	Protective Mechanism
First-level defence	Nasopharynx and upper airways receptors	Sneezing and coughing reflexes
Second-level defence	Lungs; various pulmonary receptors	Hering-Breuer and lung irritant reflexes
Third-level defence	Carotid, aortic and medullary chemoreceptors	Respiratory chemo-receptor reflexes

It appears then that both the gastrointestinal and respiratory systems are protected by three separate detection mechanisms, each succeeding level of protection being closer to the vital nervous integration centres within the brain stem.

The Area Postrema; a Closer Look at the Final Line of Defence

The area postrema (AP) is of central importance in many animal models of the vomiting mechanism as well as in man. Although ablation techniques used to study the functional role of this organ have become more precise, it is still not clear what the exact nature of the endogenous emetic stimulus(i) may be or what the role of the AP is after receiving an emetic stimulus. With respect to these questions it is useful to review what is known of the structure of the AP.

Figure 1 illustrates some of the salient morphological features of this organ. The AP is one of the number of "circumventricular organs" which are situated around the periphery of the ventricular system of the brain and which exhibit neither the structural nor physiological attributes of the blood-brain barrier.

Morphology of the Area Postrema

Its' vasculature, ependymal investment and neuronal circuitry combine to place the AP in the unique position of being able to rapidly sense humoral stimuli derived from the cardiovascular system, the ventricular cerebrospinal fluid compartment and probably the subarachnoid cerebrospinal fluid. Visceral afferents,

Fig. 1. The morphology of the area postrema. **A** Scanning electron micrograph of the area postrema (*AP*) of the cat. Note the bilobed nature of the AP as it rests on either side of the floor of the caudal end of the fourth ventricle (*IV*). Scale bar, 0.5 mm; *CC*, central canal **B** Light micrograph of a transverse section through the AP. Features of the organ which are apparent here include the highly distended perivascular spaces (*PV*) seen surrounding several small blood vessels in the parenchyma (*L*, lumen of a blood vessel); the proximity of some of these spaces to ependymal microvillous tufts (*arrow*); and the basal extensions of ependymal cells (arrowheads) which form astrocyte-like endfeet around the more dorsally positioned perivascular spaces. Scale bar, 15 μm **C** Diagrammatic representation of the AP as seen in section. The abrupt transition between normal ciliated ependyma (*ci*) and non-ciliated but microvillous (*mv*) ependyma overlying the AP is indicated. Elaborate microvillous tufts (*mvt*) often occur on apical surfaces of the AP ependymal cells and these have been seen enveloped by the macrophages (*M*) in some preparations. Occasionally supraependymal cell profiles resembling axons in section (*sen*) are seen. *Arrows* indicate the sites of poorly defined tight junction-like apical ependymal membrane specializations which have been shown to be leaky to large tracer molecules. One ependymal cell (*) extends a basal process to form part of the wall of a much-distended perivascular space (*PV*) which encloses a fenestrated (*fe*) capillary (*C*). This basal process forms one of the astrocyte-like endfeet (*af*) delimiting the wall of the perivascular space. The capillary is associated with a dorsally positioned pericyte which is enclosed within the continuous basal lamina (*bl*) of the endothelium. A second continuous basal lamina lines the wall of the perivascular space. The space itself contains abundant collagen fibres (*co*) as well as fibroblasts and profiles of vesiculated axonal varicosities. Several small neuronal somata (*ne*) are represented within the parenchyma of the *Ap* and these are closely associated with astrocyte-like glial elements (*gl*). Several axo-somatic synapses are illustrated here (*ax*) although more commonly axo-dendritic synapses (*double arrowheads*) are seen

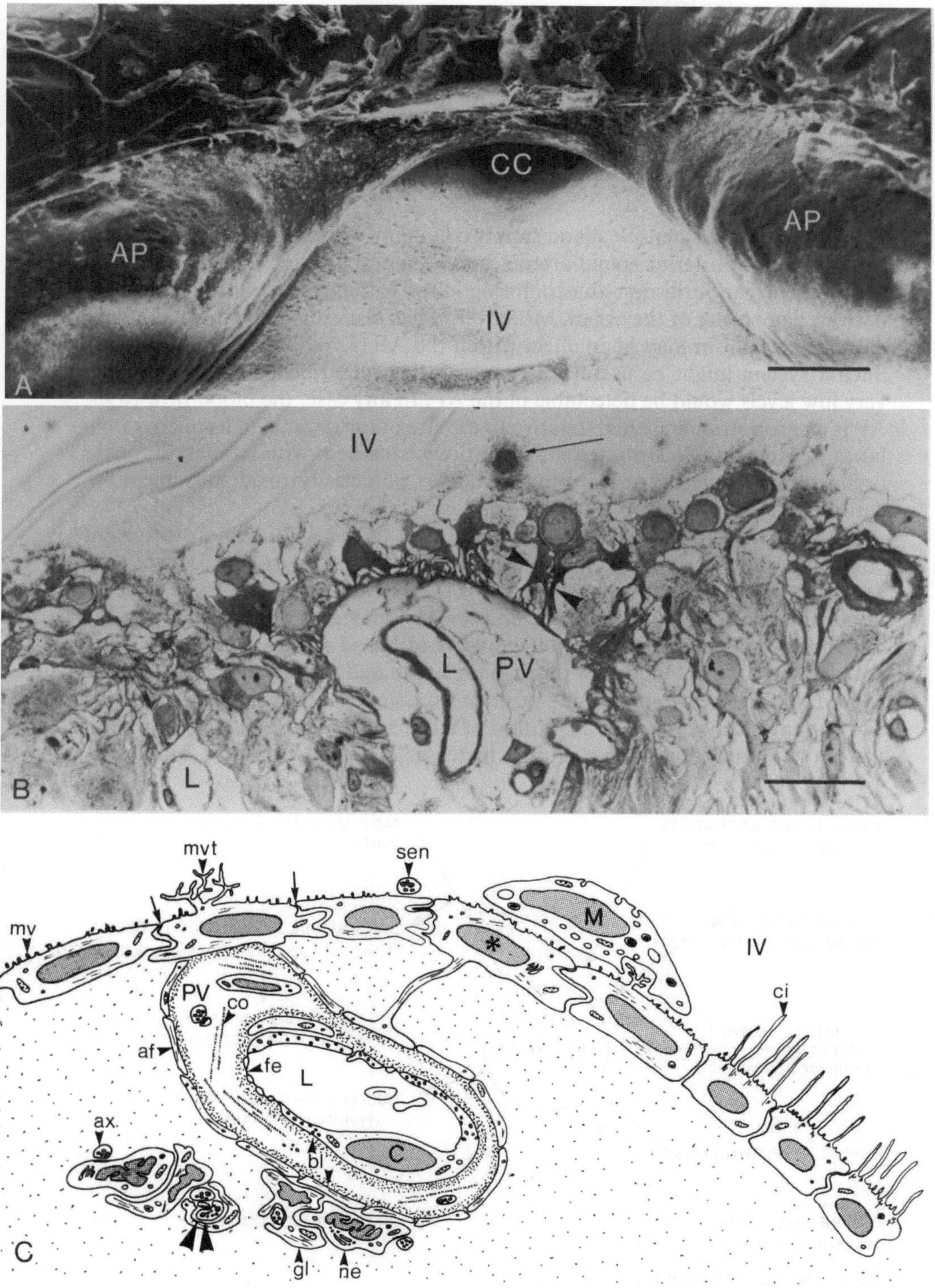
CC
AP
AP
IV
A
IV
L
PV
L
B
mvt
sen
mv
M
IV
PV
co
ci
af
fe
L
ax.
bl
C
gl
ne
C

including some from the gut, send information directly to the organ, and second order afferent information (including gustatory) is available due to the close proximity of, and neural connections with, the underlying nucleus tractus solitarii. Furthermore, efferent neurons from the AP may directly influence the dorsal motor nucleus of the vagus and others undoubtedly send information rostrally towards the pons and diencephalon to effect more central processing of visceral information. This may well be a pathway which allows gastric information to reach conscious levels.

The AP has an extensive blood supply coming from the anterior inferior cerebellar arteries, posterior spinal arteries and vertebral arteries. The blood vessels within the organ form non-anastomosing loops within the extraordinarily large perivascular spaces of the organ. Moreover, it has been suggested that a countercurrent mechanism may be in effect within the AP (Kroidl 1968; Weindl 1973). Such a system might be useful for concentrating blood-borne toxins such that very low levels would be detectable in the AP. In any case, the blood flow to the AP is disproportionately high relative to the size of the organ and its microvasculature is extensively distributed; this is entirely consistent with a function for the organ of sampling and monitoring blood for potentially toxic substances (cf. carotid and aortic bodies).

We do not yet know whether the blood flow to the AP is subject to specific control mechanisms and if so, what the nature of these might be. For example, receptors for neurotransmitter substances or putative transmitters (such as dopamine, noradrenaline, angiotensin II and opiates) may occur not on neurons, but on smooth muscle components of the walls of blood vessels of the AP. These could act to control blood flow through the organ and thereby affect its function. Indeed, recent work in man using positron emission tomography techniques has shown that intravenous administration of a dopamine agonist which only penetrates the blood-brain barrier poorly, increases blood flow in those brain areas

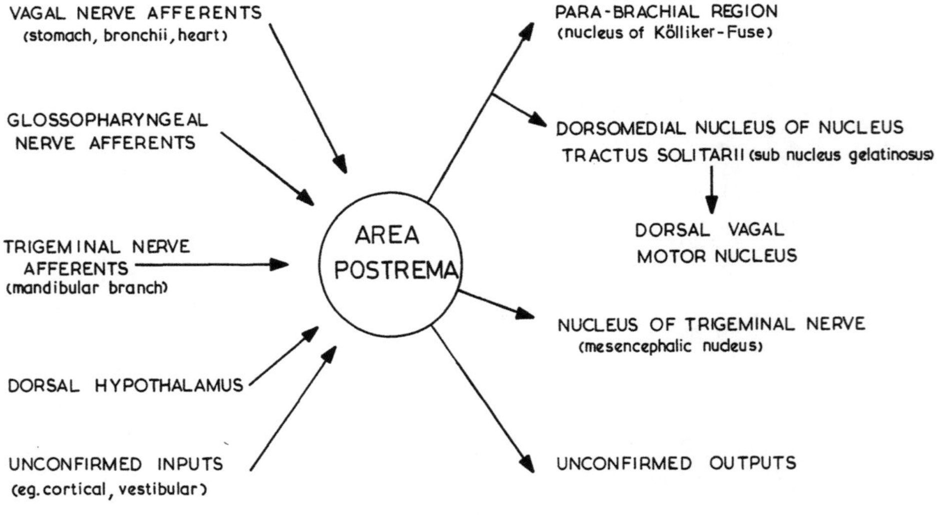

Fig. 2. The connections of the area postrema

thought to be well endowed with dopamine receptors. Subsequent administration of a dopamine (D_2) antagonist (also a poor penetrator of the blood-brain barrier) attenuated this effect (Leenders et al. 1985).

Neural elements of the AP are well positioned to sample constituents of blood or cerebrospinal fluid (see Fig. 1). The tanycyte-like ependymal cells with their extensive microvillous tufts may be important in this respect; it is, however, an open question as to whether the neuronal or the glial elements of the organ are most important in the chemoreceptive function of AP. Although the neuronal connectivity of the AP has been studied in some depth, many questions remain about the details of inputs to and outputs from the area. Figure 2 summarises our current knowledge regarding the connections of the AP. Clearly, further work is necessary to begin to establish which connections are most important for the control of nausea and vomiting.

Physiological Significance of the Area Postrema

The argument continues as to what functions the AP may subserve apart from its' role as a chemoreceptor for the emetic response. It is important to note, as Dr Borison pointed out, that ablation of AP in experimental animals results in no perceptible behavioural deficits. The AP may well act as a fine-tuning mechanism for several physiological functions (e.g. control of water intake, control of food intake, control of blood pressure and regulation of the sleep cycle) (Borison 1974). The slight decreases in mean arterial blood pressure and in body weight noted in AP-ablated rats are consistent with such a view.

The question now arises "what is the normal stimulus to the AP?". In attempting to answer this question, we might propose that from the point of view of neuronal input (e.g. from gastric vagal afferents and vestibulo-cerebellar connections), there occurs a temporal summation of neurotransmitters or their metabolites, produced as a result of excessive discordant stimuli. These substances would reach the AP by release into the blood stream and the cerebrospinal fluid. Such a mechanism might provide the necessary biochemical basis for the argument put forward to explain the cause of motion sickness (Reason 1970; Treisman 1977).

What, then, can be said about the AP as a visceral afferent fibre terminus? It is known that there are connections between the gut (principally the stomach), and areas adjacent to the AP, such as the nucleus of the tractus solitarius and the dorsal motor nucleus of the vagus. Stimulation of these gastric afferents can cause vomiting without the need to invoke any AP function, yet there are projections of these vagal afferents to the AP.

The Vomiting Centre Revisited

Very little evidence has come to light over the last 30 years to confirm the existence of a discrete group of cells to which we could ascribe the singular function of "vomiting centre" (VC). It is much more in keeping with current physiological thinking on CNS control of other functions, e.g. CVS, respiratory system, to en-

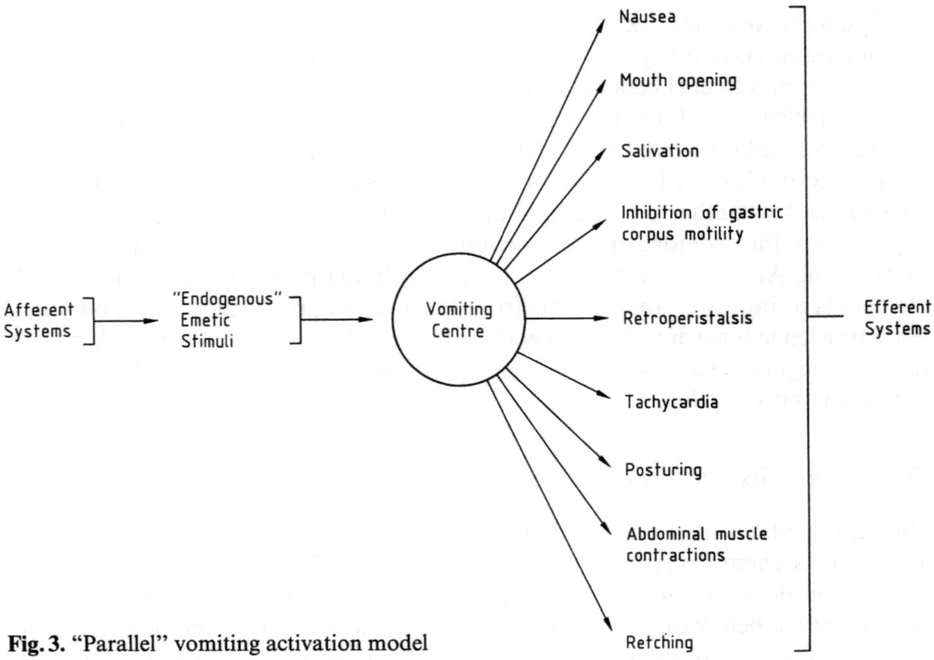

Fig. 3. "Parallel" vomiting activation model

visage a situation where control arises not from a central "black box", but is the higher expression of the integrated activity of the nuclei which ordinarily serve the separate output functions of the phenomenon in question.

The concept of the mechanism of vomiting control at the medullary level that has dominated thought for the last 30 years is represented simply in Fig. 3.

This implies two things about the "vomiting centre". One is that we should be able to locate it reliably and induce it to respond predictably and consistently to electrical stimulation with a burst of vomiting. The other is that the centre should not only be able to activate all the complex components that comprise the act of vomiting, but do so in a precise order with exact and reproducible timing. In reality, experiments to locate and stimulate the VC have been only about 50% successful and nothing has been discovered about a possible co-ordinating function of the putative VC (see, for example, Miller and Wilson 1983; Borison and Wang 1949).

Is it perhaps time, then, to consider other organisational models of the system which might explain the observations so far made and serve as a basis for experimental work to test the hypothesis. We offer an alternative possibility as summarised diagrammatically in Fig. 4. Essentially, this is a "sequential activation model" as opposed to the "parallel activation model" so long accepted as the obvious approach. It is interesting to note that the work of Openchowski and Hasko in 1887 (as cited in Hatcher's review of 1924) implied such a serial organisation of the vomiting response control mechanism.

When vomiting occurs it is an all or nothing event and there is a definite threshold for activation of the mechanism. This sequential model attempts to ex-

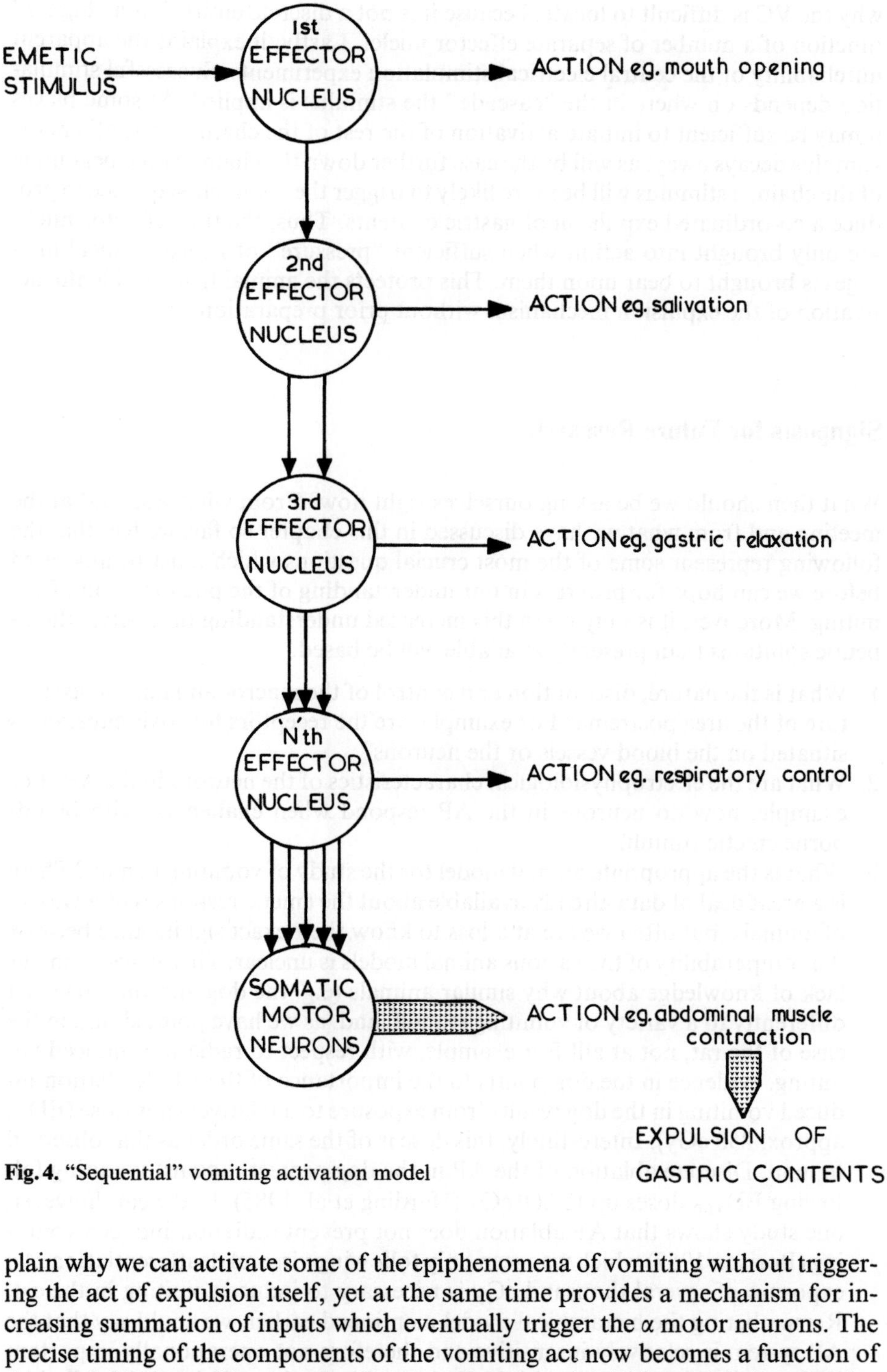

Fig. 4. "Sequential" vomiting activation model

plain why we can activate some of the epiphenomena of vomiting without triggering the act of expulsion itself, yet at the same time provides a mechanism for increasing summation of inputs which eventually trigger the α motor neurons. The precise timing of the components of the vomiting act now becomes a function of the spatial linkages between the discrete effector nuclei. This model also explains why the VC is difficult to locate: because it is not a discrete entity, but a "higher" function of a number of separate effector nuclei. Lastly, it explains the apparent unreliability of the central electrical stimulation experiments. Successful stimulation depends on where in the "cascade" the stimulus is applied. At some places it may be sufficient to initiate activation of the rest of the chain and at others the stimulus decays away, as will be the case further down the chain. At the beginning of the chain, a stimulus will be more likely to trigger the complete sequence to produce a co-ordinated expulsion of gastric contents. Thus, the final effector nuclei are only brought into action when sufficient "pressure" of neurochemical messages is brought to bear upon them. This protects the animal from accidental activation of the expulsion mechanism without prior preparation.

Signposts for Future Research

What then should we be asking ourselves right now? From what was said at the meeting and from what we have discussed in this chapter so far, we feel that the following represent some of the most crucial questions which must be answered before we can hope for progress in our understanding of the phenomenon of vomiting. Moreover, it is only upon this increased understanding that better therapeutic solutions than presently available will be based.

1. What is the nature, distribution and control of the macro- and micro-vasculature of the area postrema? For example, are the receptors for toxic substances situated on the blood vessels or the neurons?
2. What are the electrophysiological characteristics of the neurons in the AP? For example, how do neurons in the AP respond when challenged with blood-borne emetic stimuli?
3. What is the appropriate animal model for the study of vomiting in man? There is a great deal of data already available about the emetic responses of a variety of animals, but often we are at a loss to know their exact significance because the comparability of the various animal models is unclear. This arises from our lack of knowledge about why similar animals (e.g. the dog and the cat) react differently to a variety of vomiting stimuli; and, as we have pointed out, in the case of the rat, not at all! For example, with respect to radiation-induced vomiting, evidence in the dog points to the importance of the AP. Radiation-induced vomiting in the dog results from exposure to a relatively low dose (ED_{50} approx. 250 cGy). Interestingly, this dose is of the same order as that observed in man. Discrete ablation of the AP in the dog protects against vomiting following ED_{100} doses up to 800 cGy (Harding et al. 1985). In the cat, however, one study shows that AP ablation does not prevent radiation-induced vomiting (Borison 1957) which was seen only following visceral deafferentiation. An exposure of several thousand cGy is necessary to induce vomiting in the cat. Remembering the hierarchical model mentioned earlier, we could say that the AP system is relatively insensitive and therefore not operationally important in the cat in this case. It would this be interesting to learn if dogs with AP abla-

tions will vomit to the same dose necessary that ensures vomiting in the cat (5,000 cGy).

4. What is the neuroanatomical substrate subserving the subjective sensation of nausea? There was very little discussion during the meeting about the cause(s) of nausea and even less about its control. This area is an even more problematic one for research since effective anti-emetics may not prevent nausea. Indeed, it is often the case that vomiting itself actually relieves nausea.

Lastly, we must say that it may be some time before enough information becomes available to permit us to understand the process of nausea and vomiting sufficiently to enable the design of more effective and specific anti-emetics. In the interim, great value will be derived from informed controlled clinical trials. Therefore, the clinical establishment should be encouraged to perform trials with individual and combination anti-emetic therapies under the most rigorous conditions compatible with ethical practices.

References

Borison HL (1957) Site of the emetic action of X-radiation in the cat. J Comp Neurol 107:439–453

Borison HL (1974) Area postrema: chemoreceptor trigger zone for vomiting – is that all? Life Sci 14:1807–1817

Borison HL, Wang SC (1949) Functional localization of central co-ordinating mechanism for emesis in the cat. J Neurophysiol 12:305–313

Clarke GN, Davison JS (1978) Mucosal receptors in the gastric antrum and small intestine of the rat with afferent fibres in the cervical vagus. J Physiol (Lond) 284:55–68

Coil JV, Norgren R (1981) Taste aversions conditioned with intravenous copper sulphate: attenuation by ablation of the area postrema. Brain Res 212:425–433

Garcia J, Ervin FR, Koelling RA (1966) Learning with prolonged delay of reinforcement. Psychon Sci 5(3):121–122

Harding RK, Hugenholtz H, Keaney M, Kucharczyk J (1985) Discrete lesions of the area postrema abolish radiation-induced emesis in the dog. Neurosci Lett 53:95–100

Hatcher RA (1924) The mechanism of vomiting. Physiol Rev 4:479–504

Kroidl R (1968) Die arterielle und venöse Versorgung der Area Postrema der Ratte. Z Zellforsch 89:430–452

Leenders KL, Wolfson L, Gibbs JM et al. (1985) The effect of L-dopa on required cerebral blood flow and oxygen metabolism in patients with Parkinson's disease. Brain 108:171–191

Miller AD, Wilson VJ (1983) Vomiting center reanalysed: an electrical stimulation study. Brain Res 270:154–158

Reason JT (1970) Motion sickness: a special case of sensory re-arrangement. Adv Sci 26:386–393

Roper TJ (1984) A question of taste. New Sci 101(1403):31–33

Rozin P, Kalat JW (1971) Specific hungers and poison avoidance as adaptive specialisations of learning. Physiol Rev 78(6):459–486

Treisman M (1977) Motion sickness: an evolutionary hypothesis. Science 197(4302):493–495

Wang SC, Borison HL (1951) Copper sulphate emesis: a study of afferent pathways from the gastrointestinal tract. Am J Physiol 164:520–526

Weindl A (1973) Neuroendocrine aspects of circumventricular organs. In: Ganong NE, Martin L (eds) Frontiers of neuroendocrinology. Oxford University Press, New York, pp 3–32

Section II.
Clinical Aspects

Mechanisms and Treatment of Cytotoxic-Induced Nausea and Vomiting

A. L. Harris[1] and B. M. J. Cantwell[2]

Introduction

The most important subjective side-effects of cancer chemotherapy are nausea and vomiting. It is not known how such a wide variety of cytotoxic agents with different structures, modes of action and lipid solubility produce nausea and vomiting. The ability to recognise poisonous substances may have been of great evolutionary value but is a disadvantage in managing patients with cancer.

Neurological Pathways, Transmitters and Receptors Involved in Nausea and Vomiting

Chemotherapy-induced vomiting is thought to act via the chemoreceptor trigger zone (CTZ) in the area postrema (AP) in the floor of the fourth ventricle (Borison and Wang 1953). However, there are species differences and in the cat, upper abdominal de-afferentiation is necessary to abolish mustine-induced vomiting (Borison et al. 1958). Afferent pathways to the vomiting centre pass via the tractus solitarius and its nucleus and efferent pathways exit via the nucleus ambiguus and the dorsal motor nuclei of the vagus. Receptors for various different neurotransmitters are present in these pathways, e.g. muscarinic cholinergic receptors for the tractus solitarius and the nucleus ambiguus; histamine H_1 receptors for the nucleus of the tractus solitarius and the dorsal motor nucleus of the vagus, opioid receptors are also present in these two nuclei; dopamine D_2 receptors are present in the CTZ (Peroutka and Snyder 1982). Over 17 putative neurotransmitters have been described in or near the nucleus of the tractus solitarius and the AP, so that specific linking of any neurotransmitter to chemotherapy-induced nausea and vomiting at the CTZ is not possible (Borison and McCarthy 1983). However, it is possible to conclude from the anti-emetic effects of various drugs which neurotransmitters may be important in mediation nausea and vomiting at some sites in the overall reflex pathways.

Emetic Potential of Different Cytotoxic Drugs

Although all anti-cancer cytotoxic drugs can produce nausea and vomiting, there are marked differences in the emetic potency between different drugs that have

1 Department of Radiotherapy, University of Newcastle-upon-Tyne, Newcastle General Hospital, Newcastle-upon-Tyne NE4 6BE, UK
2 Wolfson Unit of Clinical Pharmacology, University of Newcastle-upon-Tyne NE1 7RU, UK

Nausea and Vomiting: Mechanisms and Treatment
Ed. by C. J. Davis, G. V. Lake-Bakaar and D. G. Grahame-Smith

otherwise similar marrow toxicity or therapeutic effect. Drugs can be ranked according to their emetic potential using average clinical doses.

It is clear that drugs such as anti-metabolites and vinca alkaloids that primarily interfere with de novo DNA synthesis or mitosis are much less emetic than drugs that primarily inhibit RNA and protein synthesis. The drugs that induce emesis in most patients are also the ones with most rapid onset of nausea and vomiting. The onset of vomiting is delayed from 1 to 12 h after an i.v. dose. It is not known why there is such a delay in onset, since peak levels of drug are present in plasma within minutes of the i.v. bolus and the CTZ is outside the blood-brain barrier.

The delay suggests that some secondary mechanisms need to be activated to produce stimulation of the CTZ. For cyclophosphamide, the time course of nausea and vomiting may be related to the need to convert it to active metabolites (Fetting et al. 1982), but peak levels of active metabolites are present within 2 h of dosing (Jardine et al. 1978) and this alone cannot explain the delayed onset.

Secondary Mechanisms for Cytotoxic Drug-Induced Vomiting

The emetic ranking by biochemical site of action of the cytotoxic drugs and the delayed onset could be accounted for if cytotoxic drug-induced vomiting depends on the activities of rapidly turning-over enzymes systems (with an intracellular

Table 1. Relationship of mode of action of drug to emetic effect and time frame of emesis

Emetic ranking	Biochemical effects	
	Special effect	General comments
1. >90%; 1–6 h Cisplatinum DTIC Cycloheximide Streptozotocin		Direct effects on proteins; methylation of protein, inhibition of protein synthesis (DNA cross-linking with cisplatinum occurs later)
2. 30–90%; 2–12 h Actinomycin D	RNA inhibition	Inhibition of RNA synthesis, directly or by intercalating and alkylating DNA (also will inhibit DNA synthesis)
CCNU, BCNU Ifosfamide Cyclophosphamide Mitomycin C	Alkylation	
Daunorubicin Adriamycin	Intercalation	
3. <30%; 6–12 h 5-Fluorouracil Ara-C Hydroxyurea 6-Mercaptopurine Methotrexate		Inhibition of de novo DNA synthesis (also have RNA effects at high doses)
Vincristine Vinblastine	Inhibition of mitosis	

half-life of approximately 6 h) that are responsible for the breakdown of a neurotransmitter (Harris 1982). In the absence of such enzymes, the neurotransmitter would increase in concentration and stimulate receptors in the CTZ. Since different agents inhibit enzymes at different points in their synthetic pathway and thus at different speeds, the involvement of enzyme systems would explain the observed differences in cytotoxic drug-induced vomiting (Table 1). Nausea and vomiting are usually observed 2–4 h after administration of cisplatin; this rapid onset of symptoms is to be expected with enzyme inhibition by heavy metals. Cycloheximide induces vomiting in man within 1 h of i.v. administration; it is an inhibitor of protein synthesis. Vomiting due to the classical alkylating agents and adriamycin usually occurs later; this delay may be due to the time it takes for enzyme levels to fall after transcription of mRNA is inhibited. The antimetabolites can cause nausea and vomiting, although they should be less likely to affect the system since de novo DNA synthesis would be at a low level. Higher doses of 5-fluorouracil, cytarabine and methotrexate can, however, interfere with RNA synthesis and it is the higher doses that cause nausea.

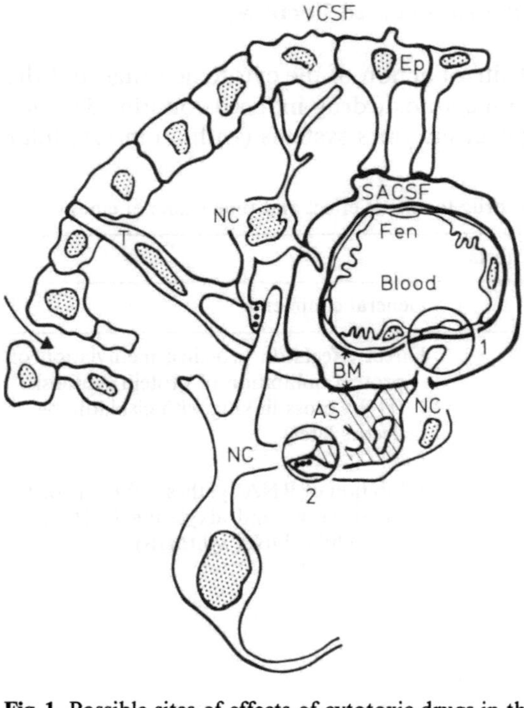

Fig. 1. Possible sites of effects of cytotoxic drugs in the area postrema (AP). Fenestrated capillaries (*fen*) are surrounded by a basement membrane (*BM*) which is surrounded by subarachnoid cerebrospinal fluid (*SACSF*). The SACSF is enclosed by a BM which separates it from the ependymal cells (*EP*), nerve cells (*NC*), astrocytes (*AS*), and tanycytes (*T*) of the AP. The ependyma has pores (→) through which ventricular csf (*VCSF*) can pass. Free nerve endings may pass into the SACSF. Astrocytes and ependyma abut on to the outer basement membrane of the subarachnoid space CSF. Potential sites of cytotoxic drug effects are: (1) the capillary endothelial wall, where nerve endings are present within the SACSF; (2) the astrocytes, which envelop synapses within the AP. (Reproduced by permission of *The Lancet*)

Two potential sites at which cytotoxic drugs may exert these delayed effects in the AP are shown in Fig. 1. One is the group of fenestrated capillaries surrounded by cerebrospinal fluid (CSF) from the subarachnoid space. The subarachnoid space surrounding the capillaries contains free nerve endings.

Another possible site consists of the astrocytes that completely surround the synapses. The astrocytes have a microtubular system that communicates with the subarachnoid space around the capillaries. Astrocytes may have a neurotransmitter role; they possess dopamine receptors, dopamine-sensitive adenylate cyclase, synthesize gamma-aminobutyric acid (GABA) and take up GABA.

Candidate Neurotransmitter Systems and the Anti-Emetic Centre Enkephalins – Emetics and Anti-emetics

Enkephalins are the most likely neurotransmitters concerned with cytotoxic drug-induced vomiting. The emetic effects of opiates are well known, and enkephalins are found in the same areas as opiate-receptor-binding activity. Areas fairly rich in opiate receptors include the AP and the subfornical organ (a small organ of unknown functions but resembling the AP in histological structure and in being outside the blood-brain barrier).

One synthetic enkephalin has produced nausea and vomiting in man. However, another that has been tested has shown none of the classic effects of morphine, perhaps because the threshold dose for prolactin and growth hormone release may be appreciably less than that for other opiate-like effects. Furthermore, the effects of morphine may be mediated by dopaminergic neurones, or they mimic the effects of dopamine release. The dopamine and enkephalin systems are closely linked and enkephalins can cause dopamine release, thus, the marginal effectiveness of anti-emetics may be due to the blocking of some secondary effects of initial enkephalin activity.

The enkephalins can be degraded by a variety of enzymes, which are found in various tissues including the brain and CSF. The inhibition of any of these enzymes would increase the concentration of enkephalin at the trigger zone. The enzymes include peptidases, enkephalinase, and arylamidases (which are inhibited by metals such as zinc and manganese). The hypothesis discussed so far has to take into account the finding by Costello and Borison that opiates have anti-emetic as well as emetic effects (Costello and Borison 1977).

Costello and Borison, who found that the anti-emetic effects of opiates were blocked by naloxone given systemically but that the emetic effects were only blocked by naloxone given directly into the cerebral ventricle, suggested that there may be an anti-emetic centre, possibly in the medullary reticular formation. They have suggested that there may be an endogenous anti-emetic tone mediated via enkephalin.

The reason opiates can produce both emetic and anti-emetic effects may be that there are at least three types of opiate receptors (u, k and δ) and possibly a dynorphin receptor. Opiate agonists can cause dopamine release via the δ receptor, but this is not blocked by naloxone. Enkephalins such as leu-enkephalin seem to bind preferentially to δ receptors. Naloxone is much less effective in inhibiting binding of enkephalins and opiates to δ receptors than to the δ receptors, which may mediate nociceptive activity. The findings of Costello and Borison may be explained by there being different types of opiate receptor at the two sites of naloxone interaction; one receptor (u receptor) is best antagonised at the anti-emetic centre, and another (δ receptor) is less well antagonised at the trigger zone.

Cytotoxic drugs crossing the blood-brain barrier at the AP may have access to the anti-emetic centre, since it has been shown that systemically administered peroxidase quickly inundates brain parenchyma adjacent to the area outside the blood-brain barrier (Broadwell and Brightman 1976). Cycloheximide inhibits enkephalin production in mouse brain. The degree of inhibition of protein synthesis associated with the fall in enkephalin production was nearly 90%. It is difficult to measure the effect in human tissue, but in susceptible tissues (leukaemic myeloblasts), a similar degree of inhibition of protein synthesis is produced within 6 h of administration of conventional doses of cytotoxic drugs. Thus, cytotoxic drugs may inhibit enkephalin production in the anti-emetic centre and potentiate effects mediated via the AP. Such a mechanism would explain why cannabinoids such as nabilone, whose effects are antagonised by naloxone, are effective anti-emetic agents.

Glial Cells and GABA

The glia may be a particularly important site for regulation of the emetic effects of anti-cancer drugs, as described earlier. GABA is usually an inhibitory neurotransmitter, and glia may play a role in limiting the effects of GABA by taking it up and metabolising it. GABA agonists such as baclofen are emetic.

The anti-emetic effects of steroids may be related to some of the above mechanisms. If the anti-emetic centre is important in providing basal control of vomiting, then steroids may help by effects on the blood-brain barrier and decreasing access of drugs to the anti-emetic centre. Steroids can also reduce leu-enkephalin release (Herz et al. 1982) and stimulate glutamine synthetase. Glutamine synthetase could divert glutamate from the GABA pathway and allow increased anti-emetic tone. Many metabolic processes in glial cells are stimulated by prostaglandins, and steroids may interfere with this mechanism. Also prostaglandins are emetic, and steroids could reduce prostaglandin release by glial cells.

These suggested testable mechanisms of activation of the CTZ or effects at an anti-emetic centre should not be confused with secondary neurotransmitter mechanisms leading from the CTZ. The latter include not only dopamine, but also histamine and acetylcholine. Although the chemoreceptors in the CTZ may be likened to olfactory or taste receptors and are responsive to many different stimuli, there is not the long latency in the latter receptors that is found in chemotherapy-induced vomiting. Benzodiazepines bind to GABA receptors (Guidotti et al. 1979), but their anti-emetic effect is more likely to be mediated via the limbic system rather than directly on the CTZ.

Anticipatory Nausea and Vomiting (Penta et al. 1983)

Whatever the biochemical mechanisms initiating nausea and vomiting, pretreatment nausea and vomiting can develop. This usually occurs after several courses of chemotherapy, may gradually become worse with time and persist after stopping chemotherapy when patients attend for follow-up. This probably represents a conditioned reflex and shows that higher cerebral or limbic stimuli may be very important in chemotherapy-induced vomiting; 1%–10% of patients develop this problem and 1%–5% of patients may refuse chemotherapy because of anticipatory nausea and vomiting.

Treatment of Nausea and Vomiting

It is important to give anti-emetics before chemotherapy and to ensure that adequate doses have been given to cover the peak onset time of vomiting. Anti-emetics may also need to be continued for 24 h or longer, depending on the cytotoxic drug. Thus, in one placebo-controlled study of high-dose metoclopramide given before and 30 min after chemotherapy, no anti-emetic effect was noted because nausea and vomiting did not start until 6–12 h after chemotherapy and lasted 24–48 h (Bui et al. 1982).

Apparently minor considerations can be important. Administration of chemotherapy to patients who are sitting is less emetic than giving therapy when they are lying flat. Effervescent drinks can reduce nausea and are better tolerated than "flat" drinks. These observations suggest that other stimuli from the gastrointestinal tract or vestibular apparatus may be synergistic with chemotherapy in producing vomiting and nausea. Patients with a previous history of motion sickness are more likely to have severe nausea and vomiting (Frytak et al. 1979).

Many different types of drugs have been used to treat nausea and vomiting, and in the past 4 years there has been a great upsurge of interest in anti-emetic trials. There are many problems in interpreting some of these trials because patients treated with different drugs, with conditioned vomiting or with different tumours have been included in the same study. Very few trials have tried to optimize drug dosage in individual patients. Therefore, the doses used in different trials can vary fivefold, even when "high dose" anti-emetics are not being evaluated. However, many randomised trials have been published and are useful in assessing the relative efficacy of the anti-emetics.

Dopamine Antagonists

Phenothiazines

Prochlorperazine is one of the most widely used anti-emetics and has become the standard to which newer anti-emetics are compared. It has been shown to be more effective than placebo, but most of the newer anti-emetics have proven superior in randomised trials. Metoclopramide, nabilone, tetrahydrocannabinoid, levona-

Table 2. Prochlorperazine (PCP) vs placebo (P)

Drug	Dose	Result	Studies
PCP	10 mg tds PO	PCP > P	Frytak et al. (1979)
PCP	5 mg tds PO	PCP > P	Moertel et al. (1963)
PCP	5 mg tds PO	PCP > P	Moertel and Reitemeier (1973)
PCP	7 mg/M^2 PO 4 dose × q 4 h	PCP = P	Orr et al. (1980)

Table 3. Prochlorperazine (PCP) vs metoclopramide (MCP)

Drug	Dose	Result	Studies
PCP	25 mg PO, 30 min pre, 3 h post chemo	MCP > PCP	Arnold et al. (1980)
MCP	20 mg PO, 30 min pre, 3 h post chemo		
PCP	20 mg i.v., 30 min pre, 3 h post chemo	MCP = PCP	Carr et al. (1982)
MCP	2 mg/kg i.v., 30 min pre + post, 3 h post		
PCP	10 mg i.m. × 6	MCP > PCP	Gralla et al. (1981)
MCP	2 mg/kg i.v. × 6		
PCP	20 mg PO × 3	MCP = PCP	Belt et al. (1981)
MCP	20 mg PO × 3		
PCP	20 mg PO × 3	MCP > PCP	Frytak and Moertel (1981)
MCP	20 mg PO × 3		

Table 4. Prochlorperazine vs cannabinoids

Drug	Dose	Result	Studies
PCP	10 mg PO q8h	NAB > PCP	Herman et al. (1979)
NAB	2 mg PO q8h		
PCP	10 mg PO q6h	NAB > PCP	Nagy et al. (1978)
NAB	2 mg PO q6h		
PCP	10 mg PO q12h	NAB > PCP	Steele et al. (1980)
NAB	2 mg PO q12h		
PCP	10 mg PO q4h	THC > PCP	McCabe et al. (1981)
THC	15 mg/M^2 q4h		
PCP	7 mg/M^2 q4h	THC > PCP	Orr et al. (1980)
THC	7 mg/M^2 q4h		
PCP	10 mg q8h	THC = PCP	Frytak et al. (1979)
THC	10 mg q8h		
PCP	5 mg, 10 mg or 15 mg q4h PO	THC = PCP	Levitt et al. (1981)
THC	5 mg, 10 mg or 15 mg q4h PO		
PCP	10 mg q4h × 3 PO	THC > PCP	Sallan et al. (1980)
THC	10 mg/M^2 q4h × 3 PO		
PCP	10 mg q4h PO	LEV > PCP	Long et al. (1982)
LEV	1 mg q4h PO		

NAB, nabilone; THC, tetrahydrocannabinol; LEV, levonantrodol

trodol, droperidol and dexamethasone were all more effective than prochlorperazine or chlorpromazine (Tables 2, 3, 4).

In some trials, doses as low as 15 mg/day prochlorperazine were used, in others 75 mg/day. There is a need to optimize present drug uses based on a knowledge of pharmacokinetics and relationships of anti-emetic drug levels to therapeutic effect. Recently, higher doses of prochlorperazine, 30 mg given i.v. every 3 h, were found to be effective in controlling cisplatin-induced emesis (Carr et al. 1984; Roche et al. 1984).

Butyrophenones

Droperidol was more effective than prochlorperazine or chlorpromazine plus promethazine (Roberts et al. 1984) and haloperidol has been shown to be anti-emetic (Neidhart et al. 1981). Both these butyropherones produce drowsiness at anti-emetic doses. The optimum dosage is not yet defined, with 1–10 mg droperidol being used i.v. in different studies (Kelley et al. 1984; Citron et al. 1984).

Metoclopramide

Metoclopramide is a procainamide derivative with peripheral and central dopamine antagonist activity. Conventional doses of metoclopramide have been compared with prochlorperazine. Although metoclopramide was more effective than prochlorperazine, neither drug was particularly effective in controlling cisplatinum-induced emesis (Table 5). The work of Gralla et al. (1981, 1982) showed that higher doses of metoclopramide (2 mg/kg 5 times a day) were more effective than placebo, prochlorperazine or cannabinoids. High-dose metoclopramide has also been used for other highly emetic regimens, including high-dose ifosfamide (5 g/m^2) (Table 6) and DTIC. The higher doses of metoclopramide were associated with diarrhoea, drowsiness and extrapyramidal side effects. The latter were commoner in young patients and were readily controlled with diazepam or anticho-

Table 5. Metoclopramide vs placebo (P), cannabinoids (THC) or lorazepam (L)

Drug	Dose	Result	Studies
MCP Placebo	20 mg q6h × 4 PO	MCP = P	Williams et al. (1980)
MCP Placebo	10 mg q8h PO	MCP = P	Morran et al. (1979)
MCP THC	2 mg/kg × 5 i.v. 10 mg/M^2 × 5 PO	MCP > THC	Gralla et al. (1982)
MCP Placebo	2 mg/kg × 5 i.v.	MCP > P	Gralla et al. (1981)
MCP Lorazepam	2 mg/kg × 5 i.v. 2.5 mg/M^2 i.v.	MCP = L	Bowcock et al. (1984)

Table 6. High dose metoclopramide or domperidone plus dexamethasone for ifosfamide-induced emesis (Harris and Cantwell 1984, unpublished results)

Ifosfamide 5 gm./M^2/24 h
High dose metoclopramide (2 mg/kg bolus 100 mg 8 hourly × 3)
51 patients
90 courses
No vomiting 48 courses (53%)
Domperidone + dexamethasone
19 patients
29 courses
No vomiting 22 courses (75%)

linergic antiparkinsonian drugs. Oral high-dose metoclopramide 2 mg/kg produces plasma levels equivalent to 2 mg/kg i.v. (Tyson et al. 1984a, b; Anthony et al. 1984). In randomised trials, it was as effective as i.v. metoclopramide (Anthony et al. 1984; Edge et al. 1984; Deppe et al. 1984). Intermediate dose i.v. metoclopramide 1 mg/kg was equivalent in activity to 2 mg/kg i.v. (Deppe et al. 1984). One pharmacokinetic evaluation suggested plasma levels of 850 mg/ml were necessary for optimal control (Meyer et al. 1984). The above studies show the need for further studies of pharmacokinetics and therapeutic efficacy.

Domperidone

Domperidone is a benzimidazole derivative and peripheral dopamine receptor blocker. It does not cross the blood-brain barrier and, hence, would be expected to have fewer extrapyramidal side effects. In adults, domperidone was equivalent to or superior to metoclopramide and domperidone was superior in children (Table 7). Domperidone was superior to placebo.

Table 7. Domperidone (DP) vs placebo (P) or metoclopramide (MCP)

Drug	Dose	Result	Studies
DP	20 mg i.v., 20 mg i.m. 3 h post chemo	DP > P	D'Souza et al. (1980)
P			
DP	16 mg i.v. × 2	DP > P	Huys (1978)
P			
DP	16 mg i.v.	DP > P	Hamers (1978)
P			
DP	1 mg/kg i.v.	DP > MCP	Swann et al. (1979)
MCP	0.5 mg/kg i.v.		
DP	12 mg i.v.	DP = MCP	D'Souza et al. (1980)
MCP	10 mg i.v.		
DP	20 mg i.v., 10 mg/*M* q6h × 4	DP > MCP	Gercovich et al. (1984)
MCP	20 mg i.v., 10 mg/*M* q6h × 4		

Table 8. Cannabinoids (THC) vs placebo (P)

Drug	Dose	Result	Studies
THC P	10 mg/M^2 q3h × 5 PO	THC > P	Chang et al. (1979)
THC P	15, 10 mg/M^2 q4h × 3 PO	THC > P	Sallan et al. (1980)
NAB P	2 mg q12h PO	NAB > P	Vincent et al. (1983)
NAB P	2 mg q6h PO	NAB > P	Nagy et al. (1978)
LEV P	1 mg q4h × 4 IM	LEV > P	Stambaugh et al. (1982)

Although higher i.v. doses of domperidone could be given because of the lack of central nervous side effects, studies with higher doses have been curtailed because of possible cardiac arrhythmias and convulsions (Weaving et al. 1984).

Cannabinoids

Three cannabinoids have undergone extensive clinical trial – Δ-9-tetrahydrocannabinol (THC), nabilone and levonantrodol. All are more effective than placebo (Table 8) and have been shown to be more effective than prochlorperazine. They do have side effects that make them less acceptable to older patients – dysphoria, dry mouth, somnolence. They have been useful in treating cisplatinum-induced emesis in younger patients. Taking the long plasma half-life of THC into account, Meyers et al. used 5.0 mg/m^2 THC 4 hourly rather than 10 mg/m^2 with significantly fewer side effects and maintained anti-emetic effect (Meyers et al. 1984).

Benzodiazepines

Lorazepam has been assessed as an anti-emetic. In one trial, the effect on memory of vomiting was more marked than the anti-emetic effect, and patients preferred lorazepam (Friedlander et al. 1981). Lorazepam has been compared with high-dose metoclopramide and was equally effective (Tyson et al. 1984a, b). Sedation is produced at anti-emetic doses.

Steroids

Dexamethasone is the most widely used steroid in anti-emetic trials. It is superior to placebo and prochlorperazine (Table 9). It is equal to intermediate dose metoclopramide (1 mg/kg × 4) in controlling cisplatinum-induced emesis. There is also a feeling of well-being and mood improvement when steroids are used as anti-

Table 9. Glucocorticoids vs placebo

Drug	Dose	Results	Studies
DEX P	10 mg i.v.	DEX > P	Cassileth et al. (1983)
DEX MCP	mg × 4 i.m. 1 mg/kg × 4 i.v.	DEX = MCP	Cognetti et al. (1984)
DEX P	20 mg × 5	DEX > P	D'Olimpio et al. (1984)
DEX PCP	20 mg i.v. 10 mg PO × 4 10 mg i.v. 10 mg PO × 4	DEX > PCP	Ettinger et al. (1984)
Prednisolone P		PRED > P	Morrow et al. (1984)

DEX, dexamethasone; P, placebo

emetics. Dexamethasone is currently being evaluated in many combination studies.

Dexamethasone added to high-dose metoclopramide was significantly better than high dose metoclopramide alone (Allan et al. 1984) and several uncontrolled studies have reported this combination to produce effective control of cisplatin-induced vomiting (McDermed et al. 1984; Strum et al. 1984). Dexamethasone plus lorazepam was also more effective than high dose metoclopramide (Gagen et al. 1984).

Other Drugs – Nortriptyline, Scopolamine, Opioids

Nortriptyline in combination with fluphenazine was superior to fluphenazine or placebos (Morran et al. 1979). There are few studies of anti-depressants as anti-emetics, but anti-depressants affect a wide range of neurotransmitters and should be further evaluated.

Transdermal scopolamine was effective in children and this mode of administration is very convenient (Sawicka and Sallan 1977). It would be worth investigating this delivery modality for any currently used anti-emetic.

Butorphanol has been studied in the ferret and had anti-emetic effects against cisplatinum, cyclophosphamide and adriamycin (Schurig et al. 1982). Opioids have not yet been assessed in man.

Combination Anti-Emetics

Several earlier trials with phenothiazines also used combinations, either with nortriptyline, lorazepam or prednisolone. Most were uncontrolled and therefore the value of combination treatment cannot be determined. Now that several moderately effective anti-emetic drugs are available and have been compared with each other and placebo, properly controlled trials can be devised.

Peroutka and Snyder (1982) discussed the neurotransmitters involved in vomiting pathways and postulated that combined blockade of histamine H_1, dopamine D_2 and muscarinic cholinergic receptors would offer optimum protection against emesis (Peroutka and Snyder 1982). Results of trials with three to five drugs have shown good activity and these studies now need to be randomised vs single agents. Combinations tested include dexamethasone, diphenhydramine and droperidol with (Sridhar et al. 1984) or without metoclopramide (Aapro et al. 1984) and a five-drug regimen of dexamethasone, diphenhydramine, diazepam, metoclopramide and either thiethylperazine (Kessler et al. 1984) or prochlorperazine (Lane et al. 1982).

Prophylaxis

Several behavioural techniques are being evaluated, including relaxation training, desensitisation and biofeedback (Cotanch 1983). However, all these techniques require extra staff and are time-consuming for patients who may be trying to lead a normal life between treatments. The best way to prevent the problem is adequate control of nausea and vomiting from the first course of chemotherapy onwards. Also, if a finite number of courses of treatment are planned, then the patient can see the end of chemotherapy. In curable tumours such as Hodgkin's disease, teratoma and acute myeloid leukaemia, maintenance chemotherapy has been shown in randomised trials to be unnecessary. In common solid tumours such as oat-cell lung cancer, 3 months chemotherapy produces similar results to 2 years treatment and in breast cancer, 6 months treatment produces the same results as treatment to relapse. Thus, appropriate planning of treatment and discussion should help to prevent the problem.

Conclusions

The choice of anti-emetics is wide and will depend on whether the patient is an in-patient or out-patient and how emetic the cytotoxic drugs are.

The majority of chemotherapy is given to out-patients and drugs such as dexamethasone, domperidone and metoclopramide orally are appropriate. Dosages of the dopamine antagonists should be adjusted individually to obtain optimum control.

Most in-patients will be receiving more emetic drugs, such as cisplatinum, ifosfamide and DTIC. Here, i.v. high-dose metoclopramide and i.v. steroids would be used. Also, the more sedative anti-emetics, such as lorazepam, droperidol, cannabinoids and chlorpromazine can be added safely.

New analogues of many of the more emetic drugs are being evaluated and should greatly reduce the problem of chemotherapy-induced vomiting (e.g. JM8, JM9 analogues of cisplatinum, and oral adriamycin analogues).

Further understanding of the neuropharmacology of cytotoxic-induced vomiting may allow more rational development of anti-emetic drugs, rather than simply selecting drugs that block apomorphine-induced vomiting.

References

Aapro M, Menoud M, Peytramann R, Alberto P (1984) Antiemetic combination dexamethasone, droperidol, diphenhydramine for DTIC or cisplatin-induced nausea and vomiting. Proc ASCO 3:C-393

Alayi JB, Torri S, Glick JH (1984) High dose oral metoclopramide: an effective antiemetic agent. Proc ASCO 3:C-424

Allan S, Cornbleet M, Warrington P, Leonard R, Smythe J (1984) Trial of dexamethasone and high dose metoclopramide vs placebo and high dose metoclopramide in cis-platinum induced emesis. Proc ASCO 3:C-345

Anthony LB, Krozely MG, Woodward NJ, Hainsworth JD, Harde KR, Brenner DE, Graco FA, Burish TG (1984) Antiemetic effect of oral vs intravenous metoclopramide in patients receiving cisplatin: a randomised double-blind study. Proc ASCO 3:C-404

Arnold DJ, Ribiero V, Bulkin W (1980) Metoclopramide vs prochlorperazine in the prevention of vomiting from diammine dichloroplatinum. Proc ASCO 21:344

Belt RJ, Rhodes J, Taylor S (1981) Antiemetic therapy for patients receiving cisdiamminedichloroplatinum (11): a randomised double-blind study comparing metoclopramide and prochlorperazine. In: Poster DS, Penta JS, Bruno S (eds) Treatment of cancer chemotherapy-induced nausea and vomiting. Masson, New York, pp 153–158

Bowcock SJ, Stockdale AD, Bolton JAR, Kang AA, Retsas S (1984) Antiemetic prophylaxis with high dose metoclopramide or lorazepam in vomiting induced by chemotherapy. Br Med J 288:1879

Borison HL, McCarthy LE (1983) Neuropharmacologic mechanisms of emesis. In: Laszlo J (ed) Antiemetics and cancer chemotherapy. Williams and Wilkins, Baltimore, pp 6–29

Borison HL, Wang SC (1953) Physiology and pharmacology of vomiting. Pharmacol Rev 5:192–230

Borison HL, Brand D, Orchard RH (1958) Emetic action of nitrogen mustard in dogs and cats. Am J Physiol 192:410–416

Broadwell RD, Brightman MW (1976) Entry of peroxidase into neurons of the central and peripheral nervous system from extracerebral and cerebral blood. J Comp Neurol 166:257–284

Bui NB, Marit G, Hoerni B (1982) High dose metoclopramide in cancer chemotherapy-induced nausea and vomiting. Cancer Treat Rep 66:2107–2108

Carr B, Pulone B, Bertrand M, Browning S, Doroshow J, Klein L, Presant C, Change FF (1982) A double-blind comparison of metoclopramide and prochlorperazine for cisplatinum-induced emesis. Proc ASCO 1:66

Carr BI, Leong L, Cecchi G, Braly P, Staples R, Blayney D, Goldberg D, Maryslin K, Bertrand M, Doroshow J (1984) Safety and efficacy of higher than conventional doses of prochlorperazine for chemotherapy-associated emesis. Proc ASCO 3:G394

Cassileth PA, Lusk EJ, Torri S, DiNubile N, Gerson SL (1983) Antiemetic efficacy of dexamethasone therapy in patients receiving cancer chemotherapy. Arch Intern Med 143:1347–1349

Chang SE, Shiling DJ, Stillman RC, Goldberg NH, Seipp CA, Barofsky I, Simon RM, Rosenberg SA (1979) Delta-9-tetrahydrocannabinol as an antiemetic in cancer patients receiving high dose methotrexate. Ann Intern Med 91:819–824

Citron ML, Johnston-Early A, Boyer MW, Krasnow SH, Priebat DA, Cohen MH (1984) Droperidol: optimal dose and time of initiation. Proc ASCO 3:C-415

Cognetti F, Pinnaro P, Carlini P, Conti EM, Cortese M, Pollera CF (1984) Randomised open cross-over trial between metoclopramide and dexamethasone for the prevention of cisplatin-induced nausea and vomiting. Eur J Cancer Clin Oncol 20:183–187

Cotanch PH (1983) Relaxation techniques as antiemetic therapy. In: Laszlo J (ed) Antiemetics and cancer chemotherapy. Williams and Wilkins, Baltimore, pp 164–175

Costello DJ, Borison HL (1977) Naloxone antagonises narcotic self-blockade of emesis in the cat. J Pharmacol Exp Ther 203:222–230

Deppe G, Smith P, Wall D, Malviya VK (1984) A comparison of the antiemetic effects of low dose and high dose metoclopramide in gynecologic cancer patients receiving cisplatin. Proc ASCO 3:C-365

D'Olimpio JT, Camacho FJ, Chandra P, Lesser M, Maldonado M, Wollner D, Wiernik PH (1984) Antiemetic treatment of patients receiving cisplatin-based chemotherapy with high dose dexamethasone v placebo. Proc ASCO 3:C-371

D'Souza DP, Reyntjens A, Thornes RD (1980) Domperidone in the prevention of nausea and vomiting induced by antineoplastic agents: a three-fold evaluation. Curr Ther Res 27:384–390

Edge S, Fundhouser W, Berman A, Seipp C, Tanner A, Wesley R, Rosenburg S, Chang A (1984) Randomised evaluation of high dose oral and intravenous metoclopramide in adriamycin/cytoxan induced emesis. Proc ASCO 3:C-378

Ettinger DS, Sheidler V, Markman M, Quaskey SA, Mellits ED (1984) Antiemetic efficacy of dexamethasone: randomised double-blind crossover study with prochlorperazine. Proc ASCO 3:C-350

Fetting JH, Grochow LB, Folstein MF, Ettinger DS, Colvin M (1982) The course of nausea and vomiting after high dose cyclophosphamide. Cancer Treat Rep 66:1487–1493

Friedlander M, Kearsley JH, Tattersall MHN (1981) Oral lorazepam to improve tolerance of cytotoxic therapy. Lancet 1:1316–1317

Frytak S, Moertel CG (1981) Management of nausea and vomiting in the cancer patient. JAMA 245:393–396

Frytak S, Moertel CG, O'Fallon JR (1979) Delta-9-tetrahydrocanna-binol as an antiemetic for patients receiving cancer chemotherapy. Ann Intern Med 91:825–830

Gagen M, Gochnour D, Young D, Gaginella T, Neidhart J (1984) Dexamethasone and lorazepam vs metoclopramide for prevention of chemotherapy-induced vomiting. Proc ASCO 3:C-396

Gercovich FG, Nahmod VE, Pirole CJ, Morgenfeld EL, Rivarola EJ (1984) A double-blind randomised cross-over study comparing domperidone to metoclopramide as an antiemetic prevention in cancer chemotherapy. Proc ASCO 3:C-344

Gralla RJ, Hri LM, Pisko SE, Squillante AE, Kelsen DP, Braun DW Jr, Bordin LA, Braun TJ, Young CW (1981) Antiemetic efficacy of high-dose metoclopramide. Randomized trials with placebo and prochlorperazine in patients with chemotherapy-induced nausea and vomiting. N Engl J Med 305:905–909

Gralla RJ, Tyson LB, Clark RA, Bordin LA, Kelsen DP, Kalman LB (1982) Antiemetic trials with high dose metoclopramide: superiority over THC and preservation of efficacy in subsequent chemotherapy courses. Proc ASCO 1:58

Guidotti A, Baraldi M, Costa E (1979) 1-4-Benzodiazepines and gamma-aminobutyric acid: pharmacological and biochemical correlates. Pharmacology 19:267–277

Harris AL (1982) Cytotoxic-therapy-induced vomiting is mediated via enkephalin pathways. Lancet 1:714–716

Hamers J (1978) Cytostatic therapy induced vomiting inhibited by domperidone: a double-blind cross-over study. Biomedicine 29:242–244

Herz A, Hoolt V, Gramsch C, Seizinger BR (1982) In: Costa E, Trabbuch M (eds) Regulatory peptides from molecular biology to function. Raven, New York, pp 51–59

Herman TS, Einhorn LH, Jones SE, Hagy C, Chester AB, Dean JC, Furnas B, Williams SD, Leigh SA, Dorr RT, Moon TE (1979) Superiority of nabilone over prochlorperazine as an antiemetic in patients receiving cancer chemotherapy. N Engl J Med 300:1295–1297

Huys J (1978) Cytostatic-associated vomiting effectively inhibited by domperidone. Cancer Chemother Pharmacol 1:215–218

Jacobs AJ, Deppe G, Cohen CJ (1980) A comparison of the antiemetic effects of droperidol and prochlorperazine in chemotherapy with cis-platinum. Gynecol Oncol 10:55–57

Jardine I, Ferselav C, Appler M et al. (1978) Quantitation by gas chromatography – chemical ionisation, mass spectrometry of cyclophosphamide, phosphoramide mustard and nornitrogen mustard in the plasma and urine of patients receiving cyclophosphamide therapy. Cancer Res 38:408

Kelley SL, Braun TJ, Rempel P, Meyer TJ, Pearlman NW (1984) Phase 1 trial of droperidol as an antiemetic in cisplatin-induced emesis. Proc ASCO 3:C-382

Kessler JF, Plezia PM, Alberts DS, Aapro MS, Surwitt EA, Graham V, Chase J (1984) Five frug antiemetic combination for prevention of chemotherapy related nausea and vomiting. Proc ASCO 3:C-406

Lane R, McGee R, Riukin S (1982) Antiemetic combinations for control of platinum induced emesis. Proc ASCO 1:65
Levitt M, Wilson A, Burman D, Fairman C, Kemel S, Krepart G, Schipper H, Weinerman B, Weinerman R (1981) Dose vs response of tetrahydrocannabinol vs prochlorperazine as chemotherapy antiemetics. Proc ASCO 22:422
Long A, Mioduszewski J, Natale R (1982) A randomised double-blind crossover comparison of the antiemetic activity of levonantrodol and prochlorperazine. Proc ASCO 1:57
McCabe M, Smith FP, Goldberg D, MacDonald J, Woolley PV, Warren R, Brodeur R, Schein PS (1981) Comparative trial of oral 9-tetrahydrocannabinol and prochlorperazine for cancer chemotherapy-related nausea and vomiting. Proc ASCO 22:416
McDermed JE, Waugh WJ, Strum SB (1984) A preliminary pharmacokinetic evaluation of high dose metoclopramide in patients receiving cisplatinum chemotherapy. Proc ASCO 3:C-409
Meyer BR, Lewin M, Drayer DE, Pasmartier M, Lonski L, Reidenberg MM (1984) Optimizing metoclopramide control of cisplatin-induced emesis. Ann Intern Med 100:393–395
Meyers F, Stanton W, Dow G, Rocchio G (1984) Reduced adverse effects with optimal antiemetic dosage schedule of delta-9-tetrahydrocannabinol. Proc ASCO 3:C-366
Moertel CG, Reitemeier RJ (1973) Controlled studies of metopimazine for the treatment of nausea and vomiting. J Clin Pharmacol 13:282–287
Moertel CG, Reitemeier RJ, Gage RS (1963) A controlled clinical evaluation of antiemetic drugs. JAMA 186:116–118
Morran C, Smith DC, Anderson DA, McArdle CS (1979) Incidence of nausea and vomiting with cytotoxic chemotherapy. A prospective randomised trial of antiemetics. Br Med J 1:1223–1224
Morrow GR, Loughner JE, Bennett JM (1984) Prevalence of nausea and vomiting and other side effects in patients receiving cytoxan, methotrexate, fluorouracil therapy with and without prednisolone. Proc ASCO 3:C-408
Nagy CM, Furnas BE, Einhorn LH, Bond WH (1978) Nabilone antiemetic crossover study in cancer chemotherapy patients. Proc AACR 19:30
Neidhart JA, Gagen M, Young D, Wilson HE (1981) Specific anti-emetics for specific cancer chemotherapeutic agents. Cancer 47:1439–1443
Orr LE; Mckernan JF, Bloome B (1980) Antiemetic effect of tetrahydrocannabinol compared with placebo and prochlorperazine in chemotherapy-associated nausea and emesis. Arch Intern Med 140:1431–1433
Penta J, Poster D, Bruno S (1983) The pharmacological treatment of nausea and vomiting caused by cancer chemotherapy – a review. In: Laszlo J (ed) Antiemetics and cancer chemotherapy. Williams and Wilkins, Baltimore, pp 81–82
Peroutka SJ, Snyder SH (1982) Antiemetics: neurotransmitter receptor binding predicts therapeutic actions. Lancet 1:658–659
Roberts WS, Wisniewski BJ, Cavanaugh D, Marsden DE (1984) Intravenous droperidol versus chlorpromazine-promethazine suppositories for the control of cis-platinum induced nausea and vomiting. Proc ASCO 3:C-419
Roche H, Hyman G, Nahas G, Stoopler M, Ellison RR (1984) Unexpected efficacy of prochlorperazine i.v. in preventing cisplatin-induced emesis. Proc ASCO 3:C-420
Sallan SE, Cronin C, Zelan M, Zinberg NE (1980) Antiemetics in patients receiving chemotherapy for cancer. A randomised comparison of delta-9-tetrahydrocannabinol and prochlorperazine. N Engl J Med 302:135–138
Sawicka J, Sallan S (1977) Transdermal therapeutic system scopolamine: prevention of vomiting associated with cancer chemotherapy. Proc ASCO 18:302
Schurig JE, Florczyk AP, Rose, WC, Bradner WT (1982) Antiemetic activity of butorphanol against cis-platinum induced emesis in ferrets and dogs. Cancer Treat Rep Oct; 66(10):1831-S
Stambaugh JE, McAdams J, Vreeland F (1982) A phase II randomised trial of the antiemetic activity of levonantrodol in cancer patients receiving chemotherapy. Proc ASCO 1:62
Sridhar KS, Donnelly E, Rosenfeld C (1984) Combination antiemetic regimen including diphenhydramine, droperidol, metoclopramide and dexamethasone in cis-platin based chemotherapy regimens. Proc ASCO 3:C-388
Steele N, Gralla RJ, Braun DW (1980) Double-blind comparison of the antiemetic effects of nabilone and prochlorperazine on chemotherapy-induced emesis. Cancer Treat Rep 64:219–224

Strum SB, McDermed JE, Streng B, McDermott N (1984) Combination metoclopramide and dexamethasone – an effective antiemetic regimen in outpatients receiving doxorubicin/cyclophosphamide chemotherapy. Proc ASCO 3:C-411

Swann IL, Thompson EN, Quershi K (1979) Domperidone or metoclopramide in preventing chemotherapeutically induced nausea and vomiting. Br Med J 2:1188

Tyson LB, Gralla RJ, Clark RA, Kris MG, Bosl GJ, Reich LM, Young CW (1984a) High dose oral metoclopramide: dose-finding efficacy and preliminary pharmacokinetic evaluation. Proc ASCO 3:C-397

Tyson LB, Stockdale AD, Bolton JAR, Kang AA, Retsas S (1984b) Antiemetic prophylaxis with high dose metoclopramide or lorazepam in vomiting induced by chemotherapy. Br Med J 288:1879

Vincent BJ, McQuiston DJ, Einhorn LH, Nagy CM, Brames MJ (1983) Review of cannabinoids and their antiemetic effectiveness. Drugs 25 [Suppl 1]:52–62

Weaving A, Bezwoda WR, Derman DP (1984) Seizures after antiemetic treatment with high dose domperidone: report of four cases. Br Med J 288:1728

Williams CJ, Bolton A, de Pemberton R, Whitehouse JM (1980) Antiemetics for patients treated with antitumour chemotherapy. Cancer Clin Trials 3:363–367

Mechanisms and Treatment of Radiation-Induced Nausea and Vomiting

R. W. YOUNG [1]

Acute Radiation Syndrome

Background

Vomiting is perhaps the most characteristic symptom of radiation sickness and often the first overt sign of radiation toxicity. The nausea and vomiting associated with radiation therapy are widely known in the lay community. In the scientific and medical community, the harmful biological effects of radiation have been the subject of concern since the late 1890s. Scientific investigation has defined the basic physiology of radiation-induced emesis, but the underlying mechanisms and the definition of an effective anti-emetic regimen remain unresolved. Within that context, this paper will review the radiobiology of emesis and the implications of the current research (see Fig. 1).

1 Biomedical Effects Directorate, Defense Nuclear Agency, Armed Forces Radiobiology Research Institute, Bethesda 20814-5145, Maryland, USA

Acute radiation symptoms

- *Nausea*
  Vomiting-upper gastro intestinal distress (UG)
- *Diarrhoea-lower gastro intestinal distress (LG)*
- *Fatiguability*
  Weakness (FW)
- *Hypotension*
  Dizziness (HY)
  Disorientation
- *Infection*
  Bleeding
  Fever (IB)
  Ulceration
- *Fluid loss*
  Electrolyte imbalance
  Headache (FL)
  Fainting
  Prostration

Fig. 1

Nausea and Vomiting: Mechanisms and Treatment
Ed. by C. J. Davis, G. V. Lake-Bakaar and D. G. Grahame-Smith

Phases of Radiation Sickness

Vomiting occurs as part of a constellation of signs and symptoms that characterize radiation sickness. The severity, scope, and course of the symptoms that occur after exposure to ionizing radiation depend on the size of the dose, dose rate, radiation quality, portion of the body irradiated, and sensitivity of the exposed individual (Langham 1967; Upton 1969). Radiation sickness is characterized by three successive phases: an initial or prodromal syndrome, a latent period, and the manifest illness (Gerstner 1958) (see Fig. 2).

Vomiting can occur as part of either the prodromal phase or the manifest phase of radiation sickness (Borison 1957; Wang et al. 1958). Prodromal emesis commonly begins within the first hour after irradiation, recurs in periodic bouts (Court-Brown 1953), and peaks in 5–8 h after exposure (Gerstner 1960). For

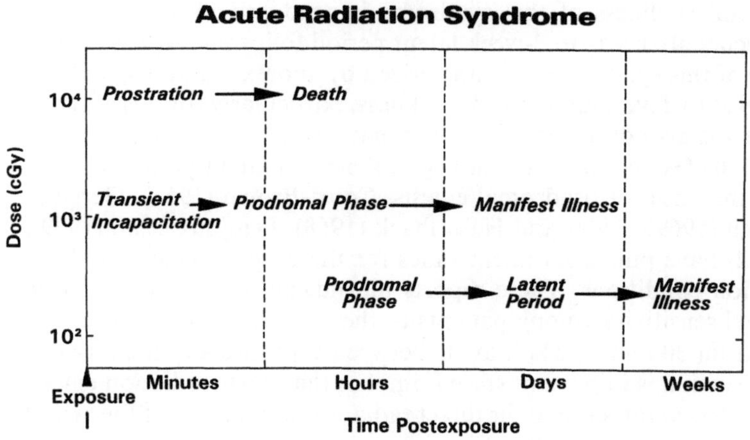

Fig. 2

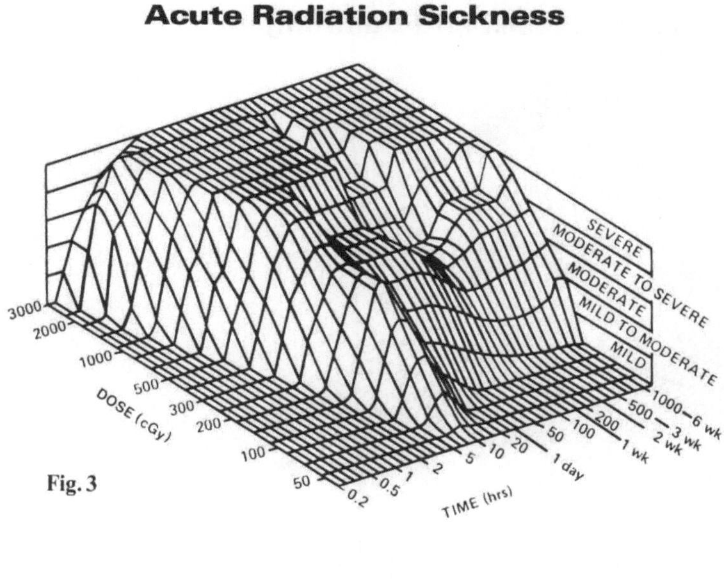

Fig. 3

doses less than 600 cGy, emesis is the first sign of radiation toxicity. The second phase of vomiting is usually delayed for several days to weeks, and is most closely associated with the terminal phase of radiation sickness. In general, the severity of nausea and vomiting is determined by the radiation dose and the resulting time course of the underlying pathology.

These parameters are summarized in Fig. 3 from the estimates contained in the present literature. Radiation sickness is usually divided into three or more syndromes that are associated with the underlying pathology of the manifest phase of the illness.

Haematopoietic Syndrome

The haematopoietic syndrome is associated with radiation doses of less than 700 cGy. The manifest illness of this syndrome is produced by bone narrow aplasia, which occurs after a 1- to 3-week latent period following exposure. The prodromal phase of this syndrome is characterized by anorexia, nausea, and vomiting accompanied by fever, fatigue and weakness, particularly for doses above 200 cGy. Vomiting is also observed during the manifest phase of the syndrome for doses above 350 cGy. As indicated in Fig. 2, there is a latent period after irradiation before the onset of prodromal emesis. Court-Brown (1953), Gerstner (1960), Lushbaugh (1969), Rider and Hasselback (1968), Danjoux et al. (1979), and Barrett (1982) have published latent times for the onset of emesis under a variety of radiation conditions. These reports all indicate that despite a wide range in individual sensitivity among patients to the same radiation exposure, a latent period of from 30 min to 2¾ h exists between exposure and the onset of emesis. This observation is important since it implies that postirradiation emesis is a second-order process rather than the direct radiation stimulation of the neural mechanisms that control vomiting (see Fig. 4).

Postirradiation Nausea and Vomiting

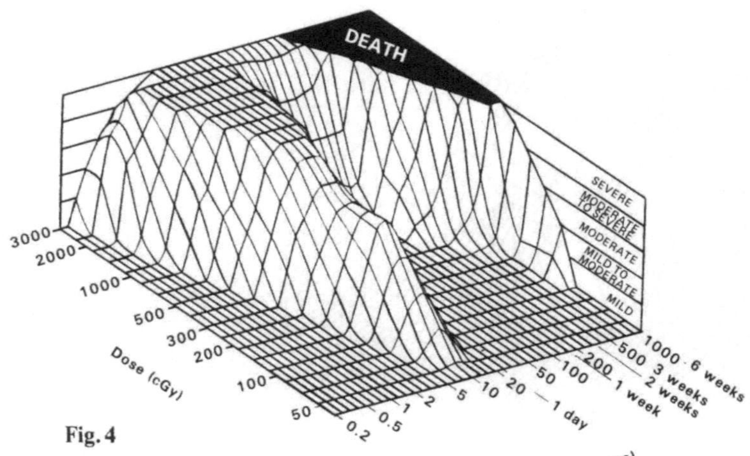

Fig. 4

Median Effective Dose for Emesis

Exposure to a total-body radiation dose of 300 cGy will cause most individuals to vomit (Lushbaugh 1969; Barrett 1982). Below 200 cGy, the incidence of emesis falls off sharply, with almost no cases of vomiting at 100 cGy or less (Lushbaugh 1969; Danjoux et al. 1979; Barrett 1982). While these general benchmarks are widely known and agreed to, establishing a dose-response relationship for radiation-induced emesis with scientific rigor has been an elusive if not impossible task because of the individual variability in the response to radiation toxicity, the lack of understanding of the mechanisms of injury, and the relative paucity of scientifically adequate human data.

Lushbaugh (1969) estimated the median effective dose (ED50) for vomiting from 504 therapy patients to be 230 cGy with 95% confidence limits between 177 and 334 cGy. This seems to be a more acceptable value than the earlier estimates that placed the ED50 below 200 cGy, in light of currently used clinical irradiation protocols that spare almost all vomiting by keeping daily exposures below 200 cGy (Barrett 1982). Above 300 cGy, the pervasiveness of radiation injury increases rapidly. For doses of 750 cGy or more, every major system in the body is affected by radiation. The degree of injury and individual susceptibility makes it almost impossible to predict with certainty the incidence of emesis above the mid-lethal dose (approximately 300 cGy). There is, however, general agreement that prodromal symptoms are produced to a greater degree by epigastric irradiation than by the same dose to other regions of the body (Gerstner 1960).

High Radiation Dose Effects

For doses of 1,000 cGy or more, the profile for radiation sickness changes. In addition to the foreshortening of survival due to the compounding of gastrointestinal and haematopoietic injury, transient functional incapacity often precedes prodromal emesis (Fig. 1). The occurrence of incapacitation is of interest to the discussion of radiation-induced emesis because vomiting is totally suppressed during incapacitation (Middleton and Young 1975). While emesis does not occur during incapacitation, monkeys that have recovered from incapacitation may subsequently vomit, although less frequently than animals that have not been incapacitated. Animals not incapacitated, but receiving the same dose as incapacitated animals, vomit as expected. The incidence of incapacitation has been shown to increase as a function of dose (Young 1973), whereas emesis in the same animals decreased as dose was increased (Middleton and Young 1975). These observations are summarized in Fig. 5. It appears that as incapacitation becomes more frequent and profound, vomiting decreases. This relationship is consistent with the observation that sedation (Barrett 1982) or depression of nervous system activity via general anaesthesia (Whitwam et al. 1978) can be an effective means of controlling vomiting in radiation therapy patients.

Since the transient incapacitation that occurs at the high doses is clearly accompanied by decreased cortical activity, reflected in shifts in the EEG to high-level slow wave activity, neural function is clearly depressed. At doses of

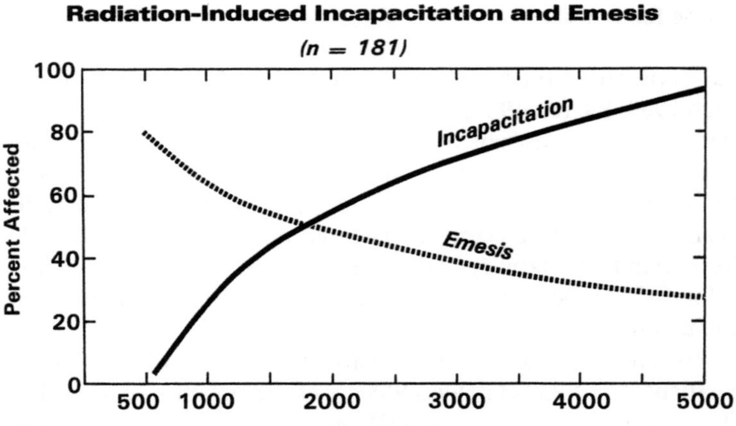

Fig. 5

10,000 cGy or more, incapacitation predominates, hypotensive shock ensues, and death follows shortly, with little vomiting evident.

Radiation-Induced Emesis

Neutron-Induced Emesis

The data that describe human emesis are almost exclusively from X-ray and gamma irradiations. Thus, the effectiveness of neutrons in causing emesis is neglected in most treatments of the subject. However, there are animal data that suggest that neutrons act differently from photons in producing radio-emesis. When monkeys were exposed to the same absorbed total-body dose in either a fission spectrum neutron or gamma field, more neutron-irradiated animals vomited and with more frequency. All of the neutron-exposed animals had two or more bouts of emesis; whereas, only one-third of the photon-irradiated subjects did (Young and Tyler 1982) (see Table 1). An even more dramatic difference was noted between monkeys given head-only irradiations of either high-energy neu-

Table 1. Frequency of vomiting for subjects who vomited as a function of *n*/Y ratio of the exposure field (Dose, 1,500–2,000 cGy)

n/Y ratio	10	3	0.4	0.1
Average dose	1,761	1,774	1,737	1,767
% Vomiting >0	40	67	56	28
Frequency ≧2	100	100	94	33
Frequency ≧3	62	80	72	17
Frequency ≧4	25	60	50	17
Frequency ≧5	12	50	28	0

Table 2. Percentage of subjects exhibiting emesis after head-only exposure to either high-energy neutrons or Bremsstrahlung

Bremsstrahlung	Dose (cGy)	High-energy neutrons
0	3,800	50
0	4,800	75
0	6,000	25
0	7,500	50
0	9,375	75

trons or matched high-energy gamma photons (young 1979) (see Table 2). In this study, neutron-irradiated animals vomited at all dose levels studied; whereas, no gamma-irradiated animals vomited. The results of these studies suggest that the more densely ionizing neutrons somehow give rise to more vomiting than do the less densely ionizing photons, particularly when the irradiations involving the head are concerned.

Central Nervous System Inflammation

Greater tissue damage and inflammation are associated with the high linear energy transfer (LET) particulate radiations (e.g. neutrons) than with low-LET radiations (e.g. gamma photons) (Rubin and Casarett 1968). As such, neutrons have been estimated to be as much as three times as potent in producing the prodromal response. Based on the work of Haymaker et al. (1958), Vogel (1959), and Wilson et al. (1960), Bond et al. (1965) have described the pathology of rapidly developing inflammatory changes in the meninges, choroid plexuses and brain after either head-only or whole-body radiation.

Following head-only irradiation of monkeys, Clemente and Holst (1954) found considerable gross evidence of oedema with herniation of the cerebellar tonsils into the foramen magnum, the existence of pressure cones, and narrowing of the sulci from X irradiation. These authors were unable to determine whether this oedema was mediated directly through the capillary wall or via cell damage to the astrocyte. From work with head irradiation of rabbits, Gerstner et al. (1956) have suggested that macromolecules broken down by radiation may increase the osmotic pressure, leading to oedema. Vasculitis, meningitis, and oedema of the choroid plexitis are clearly evident within 2–3 h after irradiation, and peak at 8 h after irradiation (Bond et al. 1965). Given oedema in the confined space of the skull, vomiting due to increased intracranial pressure certainly should be considered as a contributing factor to the radiation-induced emesis, especially in doses above 750 cGy where hypotensive ischaemia can also contribute to oedema.

Control of Radiation-Induced Inflammatory Effects

The involvement of inflammatory processes in post-irradiation vomiting is also suggested by the success of anti-inflammatory drugs in controlling emesis (see Table 3). Salazar et al. (1978), Barrett et al. (1979), and Barrett (1982) have found steroids to be effective in managing emesis in patients receiving radiation therapy. Bedikian et al. (1980) controlled headache, nausea, and vomiting with glycerol in more than 90% of patients receiving brain irradiation. Glycerol is used clinically to control cerebral oedema and lower cerebrospinal fluid pressure (Tourtellotte et al. 1972), and has been shown to protect against densely ionizing particles irradiated in vitro (Manney et al. 1963). Similarly, Stryker et al. (1979) reported that ibuprofen suppressed vomiting in patients receiving pelvic irradiations. Harding (1981) reported that the radiation-induced delay in gastric emptying in rats irradiated with 100 cGy was ameliorated by acetylsalicylic acid, indomethacin and sulindac. These latter two studies suggest that the control of inflammation in other than the head may be important in preventing post-irradiation vomiting. In general, these studies suggest that the control of inflammation and oedema may be important in the control of radio-emesis. Cairnie and Leach (1982) found dexamethasone to be a potent blocker of radiation-induced taste aversion (a conditioned taste aversion or CTA), which they used as a model for nausea and vomiting.

In the same study, insulin, domperidone, haloperidol, acetylsalicylic acid, naloxone, chlorpheniramine, cimetidine, and dimethyl sulphoxide were tested without success. In this work, dexamethasone was found to be effective against only the most sensitive mechanism of radiation-induced CTA, with other mechanisms triggered by larger radiation doses. Rabin et al. (1984) found evidence for a mechanism of radiation-induced CTA in the head that was not mediated by the area postrema (AP). In this study, it was found that the acquisition of CTA is completely disrupted by 100–200 cGy in body-only irradiated rats with AP abla-

Table 3. Anti-emetic effects of anti-inflammatory drugs

Drug	Investigator	Type of data	Effectiveness
Dexamethasone	Salazar et al. (1978)	Clinical	+
	Thomas (1983)	Clinical	+
	Cairnie and Leach (1982)	Rat (CTA)	+
Methylprednisolone	Barrett (1982)	Clinical	+
Hydrocortisone	H. R. Withers (personal communication, 1983)	Clinical	+
	Barrett et al. (1979)	Clinical	+
Glycerol	Bedikian et al. (1980)	Clinical	+
Acetylsalicylic acid	Harding (1981)	Rat (Ridge)	+
	Gralla (1979)	Dog	−
Ibuprofen	Stryker et al. (1979)	Clinical	+
Indomethacin	Harding (1981)	Rat (Ridge)	+
Sulindac	Harding (1981)	Rat (Ridge)	+

tions; whereas, after head-only irradiation the aversion was not complete. These results were interpreted as evidence for the involvement of both the AP and another mechanism in the CTA. If, in fact, increased intracranial pressure due to radiation-induced oedema plays a role in nausea and vomiting, it would provide a possible explanation for certain unaccounted – for results such as the failure of a combination of cimetidine, promethazine and thiethylperazine to offer any benefit to dogs exposed to neutrons; whereas, it produced a significant increase in the ED50 for gamma irradiations (Cordts 1982).

Physiology of Emesis

Central Nervous System

The basic physiology of radiation-induced vomiting has been established for some time. Borison and Wang (1953) have shown that the neural control mechanism for emesis is located in two distinct areas of the medulla oblongata: the chemoreceptor trigger zone (CTZ) and the vomiting centre. The vomiting or emetic centre, located in the lateral reticular formation, is the final common pathway for emesis regardless of whether the afferent input comes from the gastrointestinal tract, a cortical centre or the chemoreceptor zone. The chemoreceptor trigger zone is located in the AP on the floor of the fourth ventricle, and is activated by chemical stimuli from the blood and cerebrospinal fluid.

As such, the AP is outside the blood-brain barrier where its neurons can respond to substances within the circulation and project to the other side of the barrier to activate the motor reflexes via the emetic centre and its associated structures (Borison 1974).

Ablation of the CTZ has been shown to eliminate prodromal vomiting in the dog (Chinn and Wang 1954; Wang et al. 1958) and the monkey (Brizzee 1956). Wang et al. (1958) have further demonstrated in the dog that the vomiting associated with the manifest phase of radiation sickness is mediated through two mechanisms: the CTZ and the peripheral visceral afferent receptors of the vagus and sympathetic trunks. These conclusions are based on the observations that dogs with chronic destruction of the CTZ did not experience prodromal emesis, but did vomit later throughout their survival. Abdominal vagotomy and sympathectomy did not prevent either prodromal or delayed vomiting. However, no acute or delayed emesis was observed after irradiation in dogs with both chronic destruction of the CTZ and abdominal visceral de-afferentation.

Peripheral Emetic Stimuli

While much recent attention has focused on the role of the CTZ in vomiting, the unique sensitivity of the epigastric region to radiation must not be forgotten. Gerstner (1960) pointed out the sensitivity of patients to epigastric irradiations, and Borison (1957) found that shielding the abdomen of cats from radiation prevented vomiting; whereas, shielding the thorax and head did not. Chinn and Wang (1954) also reported that shielding of the AP did not prevent post-irradiation vomiting. While these data are often cited as indicating that radiation-induced emesis is mediated by a humoral agent released to act on the CTZ, direct stimulation fo the afferents leading to the medullary centres for vomiting cannot be ruled out. Radiation may produce emesis by local gastric irritation in much the same manner as copper sulphate. This would be consistent with the observation of Gralla et al. (1979) that acetylsalicylic acid potentiated vomiting in dogs after 800 cGy to the abdomen. Borison (1957) found that either mid-thoracic spinal cord transection or dorsal rhizotomy of the lower thoracic segments combined with supradiaphragmatic vagotomy prevented emesis in the cat. Brizzee (1956) found that supradiaphragmatic vagotomy resulted in the complete inhibition of vomiting in monkeys exposed to X-irradiation.

Despite presently unresolved interactions between the physiological components, it seems apparent that radiation-induced vomiting is mediated through the nuclei of the medulla (especially the vomiting centre and chemoreceptor trigger zone), either by activation of peripheral afferent nerves or through release of humoral factors from tissue injury, which act on the CTZ. It also appears that the extreme inflammatory processes associated with radiation injury may play a role in causing vomiting, especially at higher doses involving head irradiation.

Recent Advances

Neurotransmitter/Peptide Studies

Recent work pertinent to the mechanism and treatment of radiation-induced vomiting has examined peptide-induced emesis, the neurobiology of the chemoreceptor trigger zone, the relative effectiveness of putative anti-emetics, and the relationship between vomiting and gastric emptying. Small peptides that have been proposed as mediators of radiation-induced emesis were tested by Carpenter et al. (1984). They found that systemically administered apomorphine, angiotensin II, neurotensin, and leucine-enkephalin produced dose-dependent emesis in dogs. Chlorpromazine and AP ablation prevented the emetic response to all of these agents. Domperidone blocked the emetic response to apomorphine but not to leucine-enkephalin or angiotensin II, whereas saralasin blocked emesis from both angiotensin II and leucine-enkephalin, but not apomorphine. The results are consistent with the observation that domperidone acts as a peripheral dopamine antagonist that does not cross the blood-brain barrier. The effectiveness of chlorpromazine (also a dopamine antagonist) against emesis due to both peptides and apomorphine indicates that it has actions at some site in addition to peripheral dopamine receptors. These authors conjecture that the second site is also a dopamine receptor on the brain side of the blood-brain barrier. This raises the possibility that other anti-emetics such as phenothiazine, butyrophenone, and trimethylbenzamide derivatives also act at some central site. In work preliminary to this experiment, Carpenter (1982) reported that gastrin, substance P and VIP all appeared to produce dose-dependent emesis, but all were accompanied by indications of gastrointestinal activation, including retching, defaecation, and often an obvious increase in peristalsis. These substances were excluded from further testing as unlikely candidates, since radiation-induced emesis in animals or man is not accompanied by marked gastrointestinal activation. In this same study, mast cell preparation, cholecystokinin, secretin, 48/80, bradykinin triacetate, methionine-enkephalin, carnosine, somatostatin, and prostaglandin E1 were found to produce no emetic response (see Table 4).

In an effort to define the response of neurons in the AP to neurotransmitters and peptides, Carpenter and his co-workers (1983) determined the response of 122 neurons in the AP of anaesthetized dogs to 17 common transmitters and peptides. Excitatory responses were found to histamine, norepinephrine (noradrenaline), serotonin, dopamine, apomorphine, angiotensin II, neurotensin, leucine-enkephalin, vasoactive intestinal polypeptide, thyrotropin releasing hormone, gastrin, vasopressin and substance P. Inhibitory responses were found for norepinephrine and histamine. No response was obtained to acetylcholine, somatostatin, or cholecystokinin. All of these substances that excite the AP are emetic with the exception of serotonin and norepinephrine. None of the three agents with no excitatory effect are known to be emetic. The response of so many small neurons to so many substances suggests a common ionic or metabolic mechanism for the response. This work provides a screening matrix for thinking about circulating substances that can be released by radiation and that may serve to produce emesis by their action on the CTZ. Many of these substances have been found to be released by radiation or are postulated as possible mediators of radiation injury.

Table 4. Responses of area postrema neurons to 17 transmitters and peptides

Substance	Number of cells tested	Excitation (%)	Inhibition (%)
Glutamate	122	99	0
Serotonin (5HT)	32	66	0
Norepinephrine (noradrenaline)	15	40	40
ACh	9	0	0
Histamine	45	64	6
Dopamine	11	73	0
Apomorphine	14	86	0
Angiotensin II	68	31	0
Neurotensin	18	22	0
TRH	52	65	0
VIP	39	46	0
Gastrin	11	46	0
Substance P	23	48	0
Vasopressin	2	50	0
Leucine enkephalin	16	44	0
CCK	11	0	0
Somatostatin	10	0	0

Intestinally Derived Emetic Factor

One interesting new candidate for an intestinally derived emetic factor has recently been reported by Harding et al. (1984). These investigators found that vomiting could be induced by the intravenous injection of a side-fraction of porcine intestinal extract obtained during the purification of secretin. Dogs experienced between one and four bouts of vomiting within 3 min of injection. Pretreatment with domperidone, spiroperidol or naloxone did not block the response; whereas, discrete AP ablations did. The anatomical location of the extraction site and its dependence on the AP as a site of action make this a potentially interesting mediator of radio-emesis. Efforts of this type to identify peptides that are plausible endogenous emetic substances seem to hold promise for understanding the mechanism of radiation-induced vomiting.

Emetic Thresholds for Anti-Emetic Combinations

Mattsson and co-workers have concentrated, in a series of studies, on defining the relative potency of radiation anti-emetics, particularly combinations of drugs that have the ability to control more than one mechanism of vomiting. As a prerequisite to this body of work, these investigators undertook a systematic comparison of the monkey and the dog as models for human radio-emesis. This included determining the median effective dose (ED50) for the monkey (Mattsson and Yochmowitz 1980) and the dog (Cooper and Mattsson 1979) as baseline information. In this evaluation the ED50 was determined to be 446 ± 27 cGy for the monkey and 258 ± 38.5 cGy for the dog. Because the dog is the most similar to man

Table 5. Median effective radiation dose for dogs treated with various anti-emetics and combinations of anti-emetics (Co 60 source)

Treatment	Drug dose (mg/kg)	ED_{50} (cGy)	Target
Control	–	170	–
Promethazine	2.0	402	H1
Cimetidine	4.0	331	H2
Thiethylperazine	0.86	320	Dopamine
Naloxone	0.08	262	Opiate
(Cooper and Mattsson 1979)			
	(mg/m^2)		
Control	–	258	–
Cimetidine (C)	167.0	240	H2
Promethazine (P)	13.9	313	H1
Thiethylperazine (T)	5.6	405	Dopamine
C and P		334	H2/H1
C and T		446	H2/Dop
P and T		347	H1/Dop
C and P and T		478	H1/H2/Dop
(Mattsson et al. 1984)			

in ED50, in response to apomorphine, and in histaminase levels, it was chosen as the model for this work. In the first in this series of studies, chlorpromazine proved to be the most effective in protecting against 800 cGy. Dimenhydrinate and diphenhydramine showed less effect. Phenytoin sodium, perphenazine, WR2721, and a combination of amphetamine plus scopolamine were inactive against radio-emesis. Acetylsalicyclic acid potentiated emesis (Gralla et al. 1979). In the subsequent two studies, drug dose was held constant and midline radiation dose was varied in order to determine an ED50 for emesis for each drug tested. The results of the two (Cooper and Mattsson 1979; Mattsson et al. 1984) studies are summarized in Table 5. These studies demonstrate a technique for the quantitative comparison of drug effectiveness against emesis. The results clearly indicate that combinations of drugs can be more effective than the same drugs given singly. Further, it is clear that while these combinations of drugs raise the threshold for vomiting, once the threshold is exceeded, the syndrome is nearly as severe as that in non-treated individuals. Obviously, these dogs are not refractory to emesis despite the increased thresholds. Thus, not all of the emetic mechanisms have been suppressed. Perhaps the addition of an anti-inflammatory drug to this regime would prove effective (see Table 5).

Radiation-Induced Vomiting and Gastric Change

The interaction between radiation-induced changes in gastric motility and vomiting has been evaluated for the dog (Dubois et al. 1984) and the monkey (Dorval et al. 1984). These studies investigated the effect of Cobalt 60 irradiation on the gastric emptying of solids and liquids and the effects of the dopamine antagonist domperidone. In these carefully controlled studies, the drug and a placebo were injected blindly and in random order. Irradiation produced emesis in 9 or 10 placebo-control dogs, but in only 1 of 10 dogs pretreated with domperidone. The gastric emptying of both liquids and solids was significantly suppressed by irradiation, demonstrating that radiation-induced vomiting is accompanied by suppression of gastric emptying. However, the suppression of vomiting by domperidone did not alter the accompanying delay in gastric emptying. In the monkey, domperidone had no effect on vomiting or gastric function either before or after irradiation. These studies point up two important observations. First, in the dog at least, it appears that radiation-induced vomiting and delay in gastric emptying are independent of each other, or that the dopamine receptors involved in the two phenomena have different thresholds. The suggestion is that the dopaminergic receptors of the CTZ or the stomach are not involved in radiation-induced delay of gastric emptying in the dog. Second, in the monkey, the sensitivity of the dopamine receptors is significantly less than in the dog. Alternatively, since apomorphine does not produce vomiting in the monkey, emesis is not mediated by dopamine receptor stimulation in the monkey.

Summary

In reviewing the clinical and research data on radiation-induced nausea and vomiting, several things stand out. First, while the incidence of vomiting increases as a function of dose below the ED50 (230 cGy), it is much more difficult to predict incidence with certainty above that point. Due to the wide range of individual variation seen in the response to radiation, it is not certain that 100% of irradiated individuals would vomit at any given dose. At doses of 1,000 cGy or more, the incidence of vomiting actually decreases with dose as more individuals are incapacitated. In this case, depressed CNS function apparently suppresses emesis.

The second important observation is that neutron exposure seems to produce greater incidence and frequency of vomiting than does gamma irradiation. It is hypothesized that this stems from the greater degree of inflammation created by the high-LET radiation (neutrons). Inflammation in general appears to have been overlooked as a cause of radiation-induced emesis. The experimental evidence of oedema and the clinical success with anti-inflammatory medications suggest that increased intracranial pressure certainly contributes to post-irradiation emesis, particularly where head irradiations are involved. Reinforcing the idea that inflammation contributes to post-irradiation vomiting are the recent findings that endorphins are released by radiation and that they activate neurons in the CTZ.

Third, it must be pointed out that the role of gastric irritation in radiation-induced emesis is unclear. Because no one has been able to block vomiting during the manifest phase of radiation sickness without severing the afferents from the abdomen, it seems clear that gastrointestinal stimulation due to severe damage to the mucosa is a major triggering event for emesis during this phase of the illness. Prodromal emesis, however, is not nearly as clear. Borison's evidence from the cat (1957) indicated that the abdomen is the major initiating centre, but others (Chinn and Wang 1954; Wang et al. 1958; Brizzee 1956) have found in the dog and monkey that AP ablation can prevent early post-irradiation emesis. This fundamental difference in scientific findings illustrates the difficulty in determining the best animal model for man.

As such, one cannot ascertain from current data whether early post-irradiation vomiting is mediated by blood-borne emetic factors acting on the AP, by gastrointestinal afferent stimulation, or by a combination of these inputs.

The difficulty with animal models is further highlighted by the recent work of Dubois and his co-workers (Dubois et al. 1984; Dorval et al. 1984). In their work, domperidone, a peripheral dopamine blocker, was very effective in preventing Cobalt-60-induced vomiting in the dog but not in the monkey. Since both the control dogs and control monkeys vomited at the radiation exposure level used in these studies, it is suggested that a fundamental difference exists in the involvement of the peripheral dopamine receptor in emesis for the two species. Although the reported effectiveness of domperidone in man (Reyntjens 1979) suggests that the response of the dog is closer to that of man than the monkey, this issue needs rigorous experimental confirmation. Dubois' results also indicate that post-irradiation vomiting is independent of the changes in gastric emptying time. This observation not only provides insight into the relationship between emesis and gastric motility, but also calls into question the use of the rodent gastric emptying time model for radiation-induced emesis. As a last observation concerning modeling of post-irradiation emesis, it should be noted that the comparison of results from different radiation sources of different doses and dose rates, is hazardous experimental procedure at best. Because radiations vary in their effects on other systems for all of these factors, it should be expected that emesis would be similarly affected, particularly in view of the fact that radiation injures the tissues of so many different systems in the body.

The fourth and final observation from recent experience is that combinations of putative anti-emetics have not, to date, produced a regimen that will make individuals refractory to post-irradiation vomiting. Combinations of H1, H2 and dopamine antagonists have approximately doubled the radiation ED50 in dogs for gamma radiations, but have proved totally ineffective against neutrons. Perhaps this lack of effectiveness is due to the exclusion of anti-inflammatory drugs from the combinations tested thus far.

In any event, the procedures used in these experiments to carefully control the choice of experimental subject, control the radiation variables, and obtain quantitative estimates of the effectiveness for both individual and combinations of anti-emetics shows promise for precisely evaluating potential treatment regimens. Given the pervasiveness of radiation effects on the body and the evidence that more than one input can initiate vomiting, the systematic evaluation of other combinations of potential anti-emetics seems profitable. Aside from the basic research efforts to further define the physiology and neurochemistry of the emetic system, this work to define empirically truly effective combinations of drugs and the effort to identify radiation-released emetic factors are highly promising lines for future investigation.

References

Barrett A (1982) Total body irradiation (TBI) before bone marrow transplantation in leukaemia: a co-operative study from the European group for bone marrow transplantation. Br J Radiol 55:562–567

Barrett A, Barrett AJ, Powles RL (1979) Total body irradiation and marrow transplantation for acute leukaemia. The Royal Marsden Hospital experience. Pathol Biol 27(6):357–359

Bedikian AY, Valdivieso M, Heilbrun L (1980) Glycerol: an alternative to dexamethasone for patients receiving brain irradiation for metastatic disease. South Med J 73(9):1210–1214

Bond VP, Fliedner TM, Archambeau JO (1965) Mammalian radiation lethality. Academic, New York

Borison HK (1957) Site of emetic action of x-radiation in the cat. J Comp Neurol 197:439–553

Borison HK (1974) Area postrema: chemoreceptor trigger zone for vomiting – is that all? Life Sci 14:1807–1817

Borison HK, Wang SC (1953) Physiology and pharmacology of vomiting. Pharmacol Rev 5:193–230

Brizzee KR (1956) Effect of localized brain stem lesions and supradiaphragmatic vagotomy on X-irradiation emesis in the monkey. Am J Physiol 187:567–570

Cairnie AB, Leach KE (1982) Dexamethasone: a potent blocker for radiation-induced taste aversion in rats. Pharmacol Biochem Behav 17:305–311

Carpenter DO (1982) Peptide-induced emesis in dogs: possible relevance to radiation-induced emesis. Technical Report SAM-TR-82-28, USAF School of Aerospace Medicine

Carpenter DO, Briggs DB, Strominger N (1983) Responses of neurons of canine area postrema to neurotransmitters and peptides. Cell Mol Neurobiol 3(2):113–126
Carpenter DO, Briggs DB, Strominger N (1984) Peptide-induced emesis in dogs. Behav Brain Res 11:472–474
Chinn HI, Wang SC (1954) Locus of emetic action following irradiation. Proc Soc Exp Biol Med 85:472–474
Clemente CD, Holst EA (1954) Pathological changes in neurons, neurologia, and blood brain barrier induced by X-irradiation of heads of monkeys. AMA Arch Neurol Psychiat 71:66–79
Cooper JR, Mattsson JL (1979) Control of radiation-induced emesis with promethazine, cimetidine, thiethylperazine, or naloxone. Am J Vet Res 40:1057–1061
Cordts RE (1982) Animal-model studies of radiation-induced emesis and its control. Technical Report SAM-TR-82-26, USAF School of Aerospace Medicine
Court-Brown WM (1953) Symptomatic disturbance of the single therapeutic dose of X-rays. Br Med J I:802–804
Danjoux CE, Rider WD, Fitzpatrick PJ (1979) The acute radiation syndrome. A memorial to William Michael Court-Brown. Clin Radiol 30:581–584
Dorval ED, Mueller GP, Eng RR, Durakovic A, Conklin JJ, Dubois A (1984) Effect of ionizing radiation on gastric secretion and gastric motility in monkeys. (in press)
Dubois A, Jacobus JP, Grissom MP, Eng RR, Conklin JJ (1984) Altered gastric emptying and prevention of radiation-induced vomiting in dogs. Gastroenterology 86:444–448
Gerstner HB (1958) Acute clinical effects of penetrating nuclear radiation. JAMA 168:381–368
Gerstner HB (1960) Reaction to short term radiation in man. Annu Rev Med 11:389–302
Gerstner HB, Brooks PM, Vogel FS, Smith SA (1956) Effect of head X-irradiation in rabbits on aortic blood pressure, brain water content and cerebral histology. Radiat Res 5:318–331
Gralla FJ, Sabo JP, Hayden DW, Yochmowitz MD, Mattson JL (1979) The effect of selected drugs on first-stage radioemesis in beagle dogs. Radiat Res 78:286–295
Harding RK (1981) Ameliorating effects of drugs on radiation induced delay in gastric emptying in the rat. Radiat Res 87(2):505–506
Harding RK, McDonald TJ, Hugenholtz H, Kucharczyk J (1984) Characteristics of a new emesis-producing intestinal factor. Physiologist 27(4):279
Haymaker W, Laquer G, Nauta WJH, Pickering JE, Sloper JC, Vogel FS (1958) The effects of barium 140-Lanthanum 140 (gamma) radiation on the central nervous system and pituitary gland of macacque monkeys. J Neuropathol Exp Neurol 17:12–57
Langham WH (ed) (1967) Radiobiological factors in manned space flight. NAS-NRC Washington, DC
Lushbaugh CC (1969) Reflections on some recent progress in human radiobiology. Adv Radiat Biol 3:277–314
Maney TR, Brustad T, Tobias CA (1963) Effects of glycerol and of anoxia on the rediosensitivity of haploid yeast to densely ionizing particles. Radiat Res 18:374
Mattsson JL, Yochmowitz MG (1980) Radioation-induced emesis in monkeys. Radiat Res 82:191–199
Mattsson JL Cordts RE, Yochmowitz MG, Hardy KA (1984) Prevention of radiation emesis in dogs by combinations of drugs. Int J Radiat Oncol Biol Phys 10:1067–1072
Middleton GR, Young RW (1975) Emesis in monkeys following exposure to ionizing radiation. Aviat Space Environ Med 46(2):170–172
Rabin BM, Hunt WA, Lee J (1984) Effects of dose and of partial body ionizing radiation on taste aversion learning in rats with lesion of the area postrema. Physiol Behav 32:119–122
Reyntjens J (1979) Domperidene as an anti-emetic: Summary of research reports. Postgrad Med J 55[Suppl 1]:50–54
Rider WD, Hasselback R (1968) The symptomatic and haematological disturbance following total body radiation of 300-rad gamma-ray irradiation. In: Guidelines to radiological health Service, Publication No 999-RH-33, US Department of Health, Education and Welfare, pp 139–144
Rubin P, Casarett GW (1968) Clinical radiation pathology. Saunders, Philadelphia
Salazar OM, Rubin P, Keller B, Scarantino C (1978) Systemic (half body) radiation therapy: response to toxicity. Int J Radiat Oncol Biol Phys 4:937–950

Stryker JA, Demer LM, Mortel R (1979) Prophylactic ibuprofen administration during pelvic irradiation. Int J Radiat Oncol Biol Phys 5:2049–2052

Thomas ED, Cliff RA, Hersman J, Sanders JE, Stewart P, Buckner CD, Fefer A, McGuffin R, Smith JW, Storb R (1982) Marrow transplantation for acute nonlymphoblastic leukemia in first remission using fractionated or single-close irradiation. Int J Radiat Oncol Biol Phys 8:817–821

Tourtellotte WW, Reinglass JL, Newkirk TA (1972) Cerebral dehydration action of glycerol. Clin Pharmacol Ther 13:159–171

Upton AC (1969) Radioation injury. University of Chicago Press, Chicago

Vogel FS (1959) Changes in the fine structure of cerebellar neurons following ionizing radiation. J Neuropathol Exp Neurol 18:580–589

Wang SC, Renzi AA, Chinn HI (1958) Mechanism of emesis following X-irradiation. Am J Physiol 193(2):335–339

Whitwam JG, Morgan M, Owen JR, Goolden AWG, Spiers ASD, Goldman JM, Gordon-Smith EC (1978) General anaesthesia for high-dose total-body irradiation. Lancet 1:128–129

Wilson SG (1960) Radiation induced central nervous system death: a study of the pathologic finding in monkeys irradiated with massive doses of cobalt 60 (gamma) radiation. J Neuropathol Exp Neurol 19:195–215

Young RW (1973) The incidence of behavioral incapacitation as a function of pulsed whole-body gamma-neutron radiation dose. Annual research report ARR-7, Armed Forces Radiobiology Research Institute

Young RW (1979) Behavioral incapacitation as a function of pulsed, whole-body mixed fission-spectrum high-neutron radiation dose. Annual research report ARR-11, Armed Forces Radiobiology Research Institute

Young RW, Tyler P (1982) The incidence of postirradiation emesis as a function of radiation dose, quality, and dose rate. Radiat Res 91:389 (Meeting Abstract)

Mechanisms and Treatment of Motion Illness

J. R. R. STOTT [1]

Introduction

The question of why man develops motion sickness may be approached from several points of view. The human species is not alone in being susceptible to motion sickness: an equivalent pattern of symptoms can be induced by appropriate types of motion in a wide range of mammals (e.g. monkeys, horses, dogs, cats, seals) in birds, and even in fish. To the evolutionist, the problem is to understand what the survival value is of a pattern of symptoms evoked by certain types of motion stimulus that culminates in vomiting. From the practical point of view, it is necessary to know what the characteristics of those forms of motion are which lead to motion sickness as compared with those that do not. Such information also gives some indication of the likely neurophysiological mechanisms involved in motion sickness. From the point of view of the physiologist, interest is focussed on the mechanisms by which orientation and motion are sensed and on the sequence of events both within the central nervous system (CNS) and in other body systems that constitute the syndrome of motion sickness. Individual susceptibility in man varies widely and attempts have been made to find psychological and physiological measures that correlate with susceptibility. A further insight into the causal mechanisms of motion sickness comes from observation of the drugs which offer some protection. Such prophylactic drugs are drawn from several pharmacological groups, an indication that several neurotransmitters may be involved in the sequence of events leading to motion sickness.

Symptomatology of Motion Sickness

The sequence of symptoms and signs that constitutes motion sickness is fairly characteristic. Premonitory symptoms often include yawning or sighing, lethargy, somnolence and a loss of enthusiasm and concern for the task in hand. Increasing malaise is directed towards the epigastrium, a sensation best described as "stomach awareness", which progresses to nausea. Diversion of blood flow from the skin towards the muscles results in pallor. A feeling of warmth and a desire for cool air is often accompanied by sweating. Frontal headache and a sensation of disorientation or lightheadedness may also occur. As symptoms progress, vomiting becomes increasingly likely. In some subjects vomiting occurs early in the se-

1 Royal Air Force Institute of Aviation Medicine, Farnborough, Hampshire, UK

Nausea and Vomiting: Mechanisms and Treatment
Ed. by C. J. Davis, G. V. Lake-Bakaar and D. G. Grahame-Smith

quence of symptoms, in others malaise is severe and prolonged and vomiting is delayed.

After vomiting, there is often a temporary improvement in well being, though, with continued provocative motion, symptoms again build up and vomiting recurs.

This stage of the condition is humorously summed up by the adage, "for the first hour you fear that you will die, by the second hour you fear that you won't". However, the threat to life resulting from the repeated vomiting of sea sickness may be very real, for example aboard life rafts if rescue is delayed.

A number of objective physiological measures show changes as motion sickness develops. Plethysmography confirms the redistribution of blood flow from skin to muscles, but there are no consistent changes in pulse rate or blood pressure. There is some degree of hyperventilation and end-tidal CO_2 is reduced and arterial blood pH increased (Sinha 1968). The galvanic skin response (GSR) is a sensitive indicator of sweating, and changes can be detected on those areas of skin associated with emotive or thermal sweating (McClure et al. 1972). Oliguria during motion sickness results from an increased secretion of anti-diuretic hormone, and there is a concomitant though less dramatic increase in other pituitary hormones in particular growth hormone, prolactin and adrenocorticotrophin (Eversmann et al. 1977). These hormone responses are similar to those that occur in response to other forms of physical and surgical stress.

Characteristics of Stimuli that Provoke Motion Sickness

There are numerous circumstances in which motion sickness can be produced, many of them having been given specific names: sea sickness, car sickness, swing sickness, camel sickness, cinerama sickness, simulator sickness and most recently, space sickness. However, it is not necessary for the subject himself to be in motion in order to develop symptoms. Pilots flying fixed base aircraft simulators and members of an audience watching large screen films taken from a moving vehicle (for example, an aircraft performing aerobatics) may experience symptoms of motion sickness while they themselves remain stationary. What are the attributes of these various types of motion, either real or perceived, that lead to motion sickness? The most coherent explanation for the development of motion sickness is provided by the sensory conflict theory.

Sensory Conflict Theory

The body makes use of several sensory modalities to contribute to a perception of orientation in space and of changes in orientation that imply that the body is in motion. The suggestion that motion sickness may result from conflict between these various sensory modalities was made as long ago as 1881 by Irwin. The concept was further elaborated by Claremont (1931), Brooks (1939) and more recently by Reason (1970). Motion with respect to the outside world is sensed through the visual system; simultaneously, the vestibular system responds to angular accelerations through the semi-circular canals and to linear accelerations through the otoliths. There are, additionally, receptors in other parts of the body, cutaneous sensors, muscle spindles and tendon organs which, by sensing a changed force environment, add confirmatory evidence of motion to the afferent stimuli from the eyes and vestibular organs. In an earth-bound pedestrian environment certain rules govern the relationship between sensory inputs from these various sources. These rules are based on the normally reliable assumption that the external visual scene as a whole remains stable in space and that gravity does not change in intensity or direction. Three such rules are sufficient to define the most significant sensory interactions between the motion signals derived from vision, the semi-circular canals and the otoliths.

Rule 1. Visual-vestibular interaction. Angular motion of the head in one direction must result in angular motion of the external visual scene to the same extent in the opposite direction. A similar relationship exists for linear motion.

The contrary motion of the external visual scene is not perceived as such. Provided this rule is obeyed, the brain perceives the external world as being fixed in space. Only if the rule is violated, for example, by wearing magnifying spectacles, does the external world appear to be in motion during head movements. The expected change in the visual scene from linear motion as sensed by the otoliths is not so clear-cut as in the case of angular head motion. The amount of relative motion of an object in the visual scene depends on the distance of the object: close objects undergo large relative motion while objects at optical infinity undergo none at all.

Rule 2. Canal-otolith interaction. Rotation of the head, other than in the horizontal plane, must be accompanied by an appropriate angular change in the direction of the linear acceleration due to gravity.

This motion rule implies a fixed relationship between semi-circular canal inputs indicating head angular velocity and otolithic inputs signalling linear accelerations, primarily due to gravity. Motion of the head, for example, head tilt, must evoke equivalent sensory information from these two sources.

Rule 3. Utricle-saccule interaction. Any sustained linear acceleration is due to gravity, has an intensity of 1 g (9.81 m/s^2) and defines "downwards".

Linear accelerations are sensed by the otolith organs, those in the transverse plane principally by the utricle and in the sagittal plane by the saccule. Signals from these two component structures combine to yield sensory information about the intensity and direction of linear accelerations. Normal locomotor activities produce only transient accelerations which, over time periods of the order of 1 s, average to zero in the horizontal plane and average to 1 g, the intensity of gravity, in the vertical direction. Any sustained acceleration is therefore perceived as being due to gravity and there is an expectancy that it will remain constant in magnitude and direction. In consequence, a fixed relationship exists between utricular and saccular inputs. A sustained change in linear acceleration sensed through one component must be accompanied by an appropriate change in magnitude or direction sensed by the other component indicative of a change in head angular position within a 1-*g* environment.

The sensory conflict theory states that the common factor in the many circumstances that provoke motion sickness is the generation of patterns of sensory input that do not accord with these motion rules. Motion information obtained through one sensory modality is unsupported, or even contradicted, by the information obtained through another.

Laboratory Generated Motion Stimuli

If a subject is made to wear spectacles that produce a left to right inversion of his visual field, he has visual problems as soon as he moves his head. If he wishes to look at an object that he sees to the left of his present line of gaze, he has to turn his head to the *right*. In doing so the visual world sweeps past his gaze in the *same* direction as his angular head movement, but at twice the angular velocity. He has, in addition, severe difficulty in maintaining his balance. This pattern of sensory inputs is in direct contradiction to Motion Rule 1 and continued head movements on the part of the subject are very likely to provoke symptoms of motion sickness.

A second example of a motion sickness inducing stimulus is provided by the Coriolis or cross-coupled stimulus. If a subject rotating on a turntable at a constant velocity makes head movements in roll (bringing one ear towards the corresponding shoulder), he experiences an illusory sensation of rotation in pitch as if the chair were tipping backwards or forwards. This sensation of pitch rotation generated by signals from the semi-circular canals is not supported by signals from the otoliths to indicate change in the direction of the gravitational acceleration. Thus, Motion Rule 2 is violated. Provided turntable speeds of more than 150°/s can be generated, practically all subjects with a functioning vestibular system can eventually be made motion sick by means of cross-coupled stimulation (Graybiel et al. 1968).

A further laboratory method of inducing motion sickness is by means of low frequency vertical oscillation. A subject seated on a platform which moves up and down sinusoidally experiences a continually changing vertical force vector. Whether this stimulus generates motion sickness is very dependent on the frequency of oscillation. At frequencies of 1 Hz and above, motion sickness does not occur. As the frequency decreases below about 0.6 Hz, and perhaps maximally at 0.2 Hz, motion sickness becomes increasingly likely (O'Hanlon and McCaulay 1974). At these frequencies the accelerations generated by sinusoidal motion become more sustained and, thus, more readily interpreted as an "illegal" alteration in the intensity of gravity. It is not clear whether the development of motion sickness at low frequencies is brought about by a conflict between a previously stored estimate of gravitational intensity and that which is currently sensed, or whether there is a sensory conflict arising from the absence of an expected input from the semicircular canals. Symptoms can occur during linear vertical oscillation in subjects who keep their heads stationary and have their eyes closed. The addition of angular head movements or of whole body pitch and roll movements has not been found to potentiate the stimulus (McCaulay et al. 1976). It would appear that visual conditions play a more significant role. Motion sickness is less readily induced if the subject has a view of the stationary external world, but it is potentiated when he performs a visual search task within an enclosed visual field that moves with him.

In this circumstance the expectation that otolith-sensed linear motion will be accompanied by a visual perception of motion is not fulfilled and Motion Rule 1 is also violated.

These examples of violation of the motion rules of pedestrian man have been drawn from laboratory measures that generate motion sickness. In laboratory conditions violation of specific motion rules can be achieved in isolation. In many real-life situations in which motion sickness is provoked, multiple infringements of motion rules occur.

Vehicle-Generated Motion Stimuli

The yachtsman below decks in a small boat that pitches and rolls in a rough sea experiences conflicting sensations both in relation to the visual expectations during rotational motion (Rule 1) and also the changing force environment as the boat rides the waves (Rule 3). He is likely to feel less motion sick when on deck where a visual reference on the horizon may remove conflicting sensations in relation to angular motion of the ship, though viewing the horizon may not entirely satisfy his visual expectations in relation to linear motion. He may also find that he is less motion sick when the boat is close hauled and meeting the oncoming waves than when running before the wind. In the latter circumstances the motion of the boat feels smoother, but the wave frequency is reduced by the motion of the vessel through the water and vertical motion of the boat is at a more nauseogenic frequency. On larger ships the pitch and roll components of motion may be less prominent and the principal nauseogenic stimulus is probably the vertical excursion of the vessel (Lawther and Griffin 1981). However, the amplitude of vertical motion in the bows and stern is increased by pitching motion of the ship and is at a minimum at some point amidships (Lawther and Griffin 1981).

Violation of Rule 2, which relates the sensory inputs from semicircular canals and otoliths, is probably not an important contributor to sea sickness, but is readily produced in faster moving vehicles such as cars cornering at speed or during turns in aircraft.

When a vehicle takes a corner the occupants experience a lateral force directed radially from the centre of curvature of the vehicle's path. When this is added to the gravitational force on the vehicle occupant, the otolith organs sense these two forces at right angles to each other as a single resultant force directed at some intermediate angle. Thus, on cornering, "gravity" as perceived by the otolith organs changes in direction. This change in direction of the resultant force vector has not been brought about by head rotation so that no corresponding semi-circular canal signal is sensed.

A somewhat different situation occurs in an aircraft during a turn. Once again there is a change in the direction of "gravity" as a result of a radial force being added to the true gravitational pull. However, if the turn is correctly co-ordinated, the pilot banks the aircraft by an appropriate angle to maintain the resultant force directed downwards through the floor of the aircraft. The passengers are given the illusion that the centre of the earth remains directly down through the floor of the aircraft. The rotation in roll as the aircraft banks is sensed through the semi-circular canals, but this is unconfirmed by the otoliths which signal no corresponding change in the direction of "gravity". In a high-performance aircraft, the angle of bank during a turn is frequently greater than 60° and the amplitude of the resultant force may be many times greater than that due to gravity. In these circumstances Rule 3 is also infringed.

Adaptation to Nauseogenic Stimuli

Given a new motion environment in which terrestrial motion rules are infringed, susceptible subjects will develop motion sickness. However, with continued exposure to this environment, symptoms of motion sickness abate, typically over a period of 2–4 days. The subject regains his normal state of well-being, but may then find that on return to his previous environment motion sickness symptoms recur. This phenomenon, seen in sailors on returning to land, is termed *mal de debarquement*. Its incidence is much lower than that of the initial motion sickness and the symptoms are less severe.

In his elaboration of neural mismatch as the underlying link between the motion stimulus and the development of motion sickness, Reason (1970) puts forward a model to account for this adaptive response (Fig. 1). He postulates the existence of a neural store against which incoming patterns of motion sensation through the various sensory modalities are compared with patterns established from previous experience. Incoming sensory patterns which are not represented in the neural store evoke neural mismatch, the consequence of which is initially to provoke motion sickness. However, the continued arrival of unrepresented sensory patterns has the effect of modifying the neural store to take account of these new patterns. In consequence, after a period of time, previously unfamiliar combinations of sensory input provoke a diminishing mismatch signal and a disappearance of motion sickness. However, on return to the previous environment, the revised neural store is now inappropriate and the resulting mismatch may once again generate motion sickness.

In terms of the previously enunciated motion rules, the neural store must be regarded as the repository of such rules, and neural mismatch may be considered as synonymous with sensory conflict. However, for pedestrian man few, if any, circumstances violate these rules. Relatively few people spend extended periods in an environment where changed motion rules continuously apply. Even during long sea voyages the varying sea state determines the extent to which motion rules are violated. In adapting to ship motion in rough seas, the important modifications to the motion rules required of a passenger below deck are to reduce the dependence on visual cues for orientation and postural control in favour of a

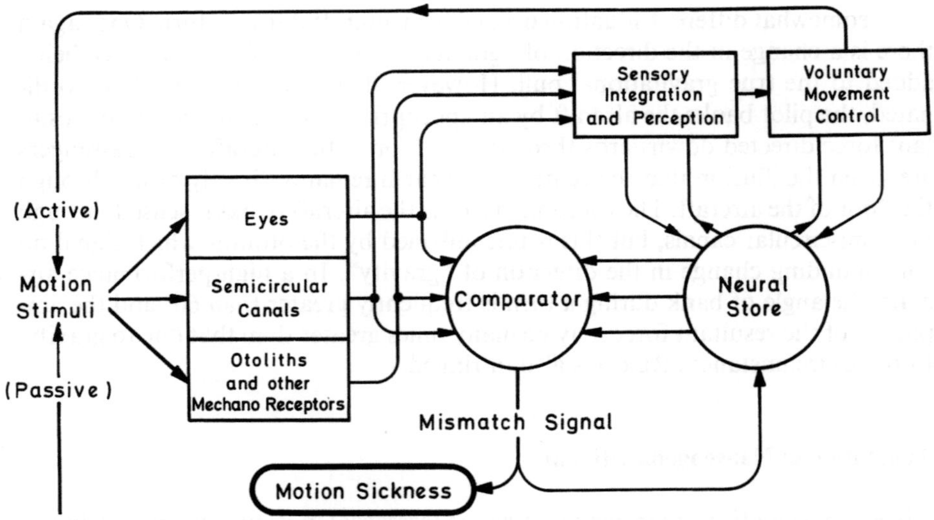

Fig. 1. The neural mismatch model of the aetiology of motion sickness. Sensory conflict between visual, semi-circular canal and otolithic motion signals in recognised by reference to a neural store of legitimate sensory patterns. Any resulting mismatch has the immediate effect of provoking symptoms of motion sickness and the longer term effect of altering the neural store to take account of previously conflicting signals and, hence, to reduce the degree of mismatch

heightened perception of, and dependence on, otolithic and somaesthetic sensors, appropriately termed "getting one's sea legs". There must also develop an expectancy of continually changing gravitational intensity as evidenced by the illusory heaving motion of terra firma on return to land when that expectancy is no longer fulfilled.

One environment in which new motion rules consistently apply is provided by space flight. Motion sickness has complicated space missions almost from the start, though the incidence was less when spacecraft were small and restricted the movements of the occupants. In the zero-g environment of space, signals from the otolith organs are generated only in response to the short duration linear accelerations produced by movement of the astronaut within the spacecraft. The otolithic input becomes valueless for postural control and for a perception of "up" and "down" – a function which must be subserved entirely by vision. Perhaps most significant for the development of motion sickness is the fact that angular head movements are no longer accompanied by an appropriate otolithic signal to indicate a change in the direction of gravity. This represents a sensory conflict between canals and otoliths which may well be the predominant conflict in the production of space sickness (Benson 1977). The symptoms of space sickness are liable to develop during the first few hours of flight precipitated by activity and, particularly, by head movements (Homick 1979). Symptoms abate over the course of 3–4 days after which the astronaut can, in general, indulge in unlimited activity without developing symptoms (Oman et al. 1984). Once adapted to the environment of space, astronauts have been found to be immune to cross-coupled stimulation – the laboratory method of generating a canal-otolithic conflict. This is perhaps to be expected since in space there is no otolithic signal to conflict with the sensation of rotation induced by the cross-coupled stimulus. However, on return to earth, this immunity to cross-coupled stimulation persists and pre-flight levels of susceptibility are only gradually regained over several weeks. This observation lends weight to the idea that adaptation to the motion environment of space entails a discounting of otolithic inputs and an increased reliance on visual cues for the purposes of orientation (Young et al. 1984). Adaptation to the shipborne environment involves an opposite process – that of discounting visual cues in favour of otolithic and somaesthetic inputs. Thus, astronauts might not be expected to be immune to sea sickness on their return to earth and indeed some have developed sea sickness in the capsule on splash down or aboard the recovery vessel (Homick 1979).

In most situations that provoke motion sickness, an individual's exposure is short compared with the 3–4 day adaptation period that occurs on a sea voyage or in space. Such exposures may be a few minutes on a fairground motion amusement, a 1–2 h sortie in a high-performance jet aircraft or a road journey or sea crossing of a few hours.

Nonetheless, adaptation to these intermittent provocative stimuli does occur with repeated exposure, though perhaps not to the complete extent that occurs with space sickness. The degree of adaptation achieved depends on the individual, in particular his underlying susceptibility, the rate of adaptation during exposure to the provocative stimulus and the rate at which adaptation is lost during the intervening periods when the normal terrestrial motion rules apply.

Pathologically Induced Motion Stimuli

The concept that abnormal patterns of sensory input from motion sensors are responsible for the generation of motion sickness leads to an explanation of the nausea and vomiting that is a feature of labyrinthitis and of Meniere's disease and which almost invariably follows unilateral labyrinthectomy. In all these conditions vertigo is the result of an abnormal pattern of neuronal discharge from one labyrinth. This neuronal input is unsupported either by visual signals of motion or by an appropriate pattern of otolithic signals, and a state of sensory conflict is therefore present.

Primary afferent neurons from the semi-circular canals exhibit a resting discharge of about 80 spikes/s. Following an angular motion stimulus, an increased discharge rate from one semi-circular canal is accompanied by a decrease in the discharge rate from the corresponding canal on the contralateral side. The absence of a resting discharge that follows ablation of one labyrinth, therefore, represents a strong rotatory signal from the non-functioning semi-circular canals on that side. Thus, in addition to visual and otolithic conflict, there is also generated an inter-labyrinthine conflict. The time course of the nausea and vomiting that follows unilateral labyrinthectomy is similar to the 2- to 4-day adaptation period required by a susceptible subject in an abnormal motion environment.

Evolutionary Aspects of Motion Sickness

There is an apparent lack of purpose in a sequence of symptoms that leads to vomiting in response to an unfamiliar motion environment. It has been argued that motion sickness represents an evolutionary anomaly that has only come to light as a consequence of modern forms of transport (Oman 1982). However, the fact that a large number of animal species can be made motion sick justifies the quest for an evolutionary origin of the phenomenon. Triesman (1977) has proposed that vomiting is an appropriate response with survival value which is elicited by an inappropriate stimulus, namely, discordant motion signals. He points out the functional importance to any animal of the sensory information which enables it to organise its movements within a spatial framework. Because this sensory information is derived from multiple sources, notably vision, the vestibular system, and somatic proprioceptive sensors, there is a continuous need to co-ordinate and to detect incongruities in the pattern of information from these various sources. Only when this information is coherent can the animal generate an appropriate neuromuscular response either to redirect gaze or to make postural adjustments. Because this CNS activity is likely to be highly integrated and continuously monitored it might also be readily affected by changes in neural function brought about by the ingestion of neurotoxic substances. If this were so, then the evolution of an emetic response to sensory disorganisation might have survival value. The fact that certain motion environments engender a similar loss of coherence in motion-related sensory inputs and so trigger the same emetic response now appears as an unfortunate coincidence.

One observation that may be of relevance to this theory is that the emetic response of dogs to certain substances, in particular lobeline, L-dopa and nicotine, is reduced following labyrinthectomy, a procedure which also abolishes motion sickness. The emetic response to apomorphine and pilocarpine is, however, unchanged (Money and Cheung 1983).

Motion Sickness Susceptibility

While the sensory conflict theory is successful in accounting for those types of motion that generate motion sickness, a similar success cannot be claimed in the search for physiological or psychological correlates that account for the wide differences in susceptibility between individuals.

Susceptibility to motion sickness is in itself a complex attribute. A subject's innate sensitivity to sensory mismatch may differ from one type of motion to another, and in consequence it is only broadly true that subjects susceptible to one environment will experience motion sickness in another. Subjects also vary in their rates of adaptation when they encounter an abnormal motion environment.

The innate sensitivity to mismatch has been termed receptivity (Reason 1968). Subjects who on a questionnaire study were highly susceptible were found, when making magnitude estimations of other sensory modalities such as loudness, brightness, or weight, to generate steeper than normal magnitude estimation curves.

Subjects also vary in their rates of adaptation when they encounter an abnormal motion environment. One means of assessing this attribute has been to measure the amount of exposure to a cross-coupled stimulus before its illusory components cease to be evident (Reason and Graybiel 1969). This measure, termed adaptability, also correlated with motion sickness susceptibility. However, experimental evidence is lacking as to whether it also reflects true adaptability: that is, the longer term rate at which tolerance is acquired to repeated exposures of a nauseogenic stimulus. It was shown that these attributes of receptivity and adaptability, as defined by Reason, were largely independent of each other. Adaptability was found to be the more significant contributor to susceptibility.

A further variable factor in an individual's overall susceptibility is the extent to which, having adapted to a new motion environment, adaptation is retained during intervals between exposure. This factor has not been extensively explored.

Astronauts have been shown to lose over a period of about 12 weeks the adaptation to cross-coupled stimulation conferred by adaptation to weightlessness (Graybiel et al. 1974). On the other hand, aircrew who have undergone desensitisation treatment and have acquired an adaptation specifically to cross-coupled stimulation have been observed to retain some degree of adaptation when re-exposed to the stimulus after an interval of a year or more (Stott and Bagshaw 1984).

Although a functioning vestibular system is essential to the development of motion sickness, tests of vestibular function give little useful indication of susceptibility. When subjects in a rotating chair undergo an abrupt stop, the intensity of the resulting nystagmus is not correlated with motion sickness susceptibility (Benson 1973). Subjects given a series of such post-rotational tests show a decline of nystagmus output with each succeeding test. The rate of this decline shows some correlation with questionnaire-assessed susceptibility, the decline being less marked in more sesceptible subjects (Hill 1984).

Sensation cupulometry measures the duration of the sensation of rotation that follows a stopping stimulus from various speeds of rotation. Initial reports that this test gave an indication of susceptibility were not confirmed (Dobie 1974). Susceptible subjects tend to be more introvert (Kottenhoff and Lindahl 1960) and possibly more neurotic (Wilding and Meddis 1972), but correlations are weak. The search for tests indicative of susceptibility has been stimulated by the desire to predict failures in training attributable to motion sickness among aircrew or astronaut applicants. It might be expected that a simple motion-sickness-provoking laboratory test would be sufficient to identify such subjects. A number of tests have been devised (reviewed in Reason and Brand 1975). However, although broadly predictive, they failed to pick up all those who had to withdraw from training on account of motion sickness and would have excluded others who successfully completed training. Essentially such tests fail to take account of adaptation in the course of training, and any screening test to take account of this variable is likely to be time-consuming. At present, in aircrew selection, no account is taken of a history of motion sickness and no selection test is used.

Because of a wider choice of potential applicants for astronaut training, those with a high susceptibility, as indicated by laboratory tests using cross-coupled stimuli, can be excluded. However, there is as yet no test that will predict those who will be affected by space sickness.

Neurophysiology of Motion Sickness

Knowledge of the neural pathways involved in the generation of motion sickness is far from complete. The suggestion that a functioning vestibular system is necessary for the development of motion sickness was first made by Irwin in 1881. This view was strengthened by observations on deaf-mutes by James (1882) and Minor (1896). They found that about one-third of deaf-mutes experienced no dizziness after being spun on a rotating chair, implying a diminished or absent vestibular function. Despite some subjects in this group having had ample opportunity, none had ever experienced seasickness. Since that time, many further observations have been made to indicate the total immunity to all forms of motion sickness enjoyed by those who lack vestibular function. Vision is essential in the generation of some forms of motion sickness but, these apart, blind people probably have similar motion sickness susceptibilities to those who are sighted (Graybiel 1970).

Figure 2 summarises the principal intracerebral structures that are involved in the development of motion sickness. Primary afferent neurons from the vestibular apparatus run to the vestibular nuclei and also project as mossy fibres to the cerebellar cortex. From the vestibular nuclei, second order neurons project to

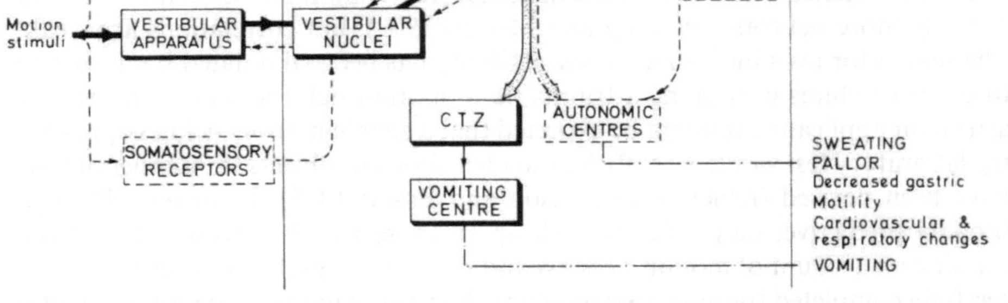

Fig. 2. Diagram of the principal neural interconnections involved in motion sickness. The *heavily outlined boxes* indicate structures that are essential to the development of vomiting in response to provocative motion. Motion-induced sickness may, however, occur via neural pathways that do not involve the CTZ

the oculomotor nuclei via the medial longitudinal fasciculus, to the reticular formation, and via the inferior peduncle to the cerebellar cortex. The involvement of regions of the cerebellum in the generation of motion sickness was established in dogs by Bard et al. (1947) and by Wang and Chinn (1956). These latter workers showed that dogs were rendered immune to motion sickness by removal of the cerebellar nodulus and uvula, the region of the cerebellum that receives most of the projections from the vestibular nuclei.

Previous studies in dogs (Wang and Borison 1952; Wang and Chinn 1954) had elucidated the functional relationship between the chemoreceptor trigger zone (CTZ) in the area postrema (AP) of the medulla and the vomiting centre in the region of the fasciculus solitarius. An intact CTZ was shown to be necessary for the vomiting response both to apomorphine and to swing-induced motion sickness, though the motion sickness response in two of 12 dogs was unaffected by ablation of the CTZ perhaps indicative of some alternative pathway. However, whereas motion-induced vomiting required an intact vestibular apparatus and vestibulo-cerebellum, surgical removal of these structures did not prevent vomiting following apomorphine administration.

Other regions of the brain are involved during the development of motion sickness. The stress hormone response is mediated through the hypothalamus and pituitary, and autonomic activity is responsible for pallor, sweating, and changes in gastric motility. Cortical structures are also implicated in motion sickness. In experimentally induced motion sickness, subjects who performed a mental arithmetic task delayed the onset of motion sickness compared with those who were asked to report their perception of body position (Guedry 1970). However, motion sickness can still be elicited in decorticate animals (Bard et al. 1947) and air sickness is thought to have been observed in a decorticate man (Doig et al. 1953). Ablation studies have shown that neither the hypothalamus nor the pituitary is essential to the generation of motion sickness (Money and Wood 1970).

Autonomic innervation of the gut, although essential to vomiting induced by copper sulphate, a peripherally acting emetic agent, is not necessary for the development of motion sickness. Vagotomised and sympathectomised dogs remain susceptible to motion-induced vomiting, and also to centrally acting emetic substances (Wang et al. 1957).

The sensory conflict theory, in ascribing motion sickness to a lack of accord between motion signals derived from different sensory modalities, focuses attention on those areas of the brain where these sensory modalities converge. Two regions are of particular significance, the vestibular nuclei and the flocculus, nodulus and uvula of the cerebellum.

Cells of second-order neurons in the vestibular nuclei can be modulated in their firing rate not only by vestibular stimuli, but also by optokinetic inputs and by joint movement. Visual and vestibular interactions have been more extensively studied. The semi-circular canals respond to angular accelerations of the head, but their dynamics dictate that the signal they generate in primary afferent neurons is one of head angular velocity. Following a step change of head angular velocity, a new firing rate is established which decays toward its resting level with a time constant of 5.7 s as measured in the monkey (Fernandez and Goldberg 1971). This same velocity signal can be detected in specific second order neurons of the vestibular nuclei though its time constant of decay is some three times longer. In addition, a similar direction-specific response in these neurons can be generated in the absence of rotation by optokinetic stimuli that involve the whole visual field (Waespe and Henn 1977). Unlike the response to the vestibular stimulus, this response is sustained, but saturates at stimulus velocities of about 60°/s. When visual and vestibular motion stimuli are presented simultaneously and are concordant, a complementary response is seen in the second order neuron which, unlike the purely vestibular response, accurately follows the angular velocity at low rates of acceleration and operates to higher velocities than with the visual stimulus alone. If, however, visual and vestibular stimuli conflict, the signal generated in the second order neurons is reduced in amplitude and decays rapidly (Waespe and Henn 1978).

Another region of interaction between visual and vestibular inputs is the flocculus of the cerebellum. Recordings of single Purkinje cell activity in the flocculus indicate that modulation of firing rate was greatest during the pursuit of a visual target in the absence of head movement, of alternatively when vestibular-induced eye movements had to be suppressed in order to track a visual target that moved with the head. Modulation of activity was least when eye movements were compensatory for head movements during fixation of a space stable visual target (Lisberger and Fuchs 1978). The Purkinje cell output can, thus, be considered as a conflict signal between ocular and vestibular signals of motion.

There is also evidence that the modification of vestibulo-ocular reflex that can be induced by prolonged left-right visual image reversal is a function of the input to the flocculus derived from the inferior olive (Ito and Miyashita 1975). Thus, in this region of the cerebellum are to be found signals representing motion cue conflict and also pathways that bring about an adaptive response to persistently conflicting sensory cues. Both of these features are essential requirements of the sensory conflict model of motion sickness. The principal outputs of the flocculus project to eye movement control areas within the vestibular nuclei, but there are other projections which are not apparently related to this function (Dufossé et al. 1977).

The neurophysiological functions of the nodulus and uvula of the cerebellum, regions which in dogs are necessary for the generation of motion sickness, are less fully elucidated. These areas receive primary neurons from the vestibular nerve and secondary neurons from the vestibular nuclei as well as fibres of visual origin and are, therefore, also likely to be involved in the interactions between different sources of motion information.

The subsequent neural pathways that bring about motion sickness remain unclear. Whereas, the sensory integration of visual and vestibular motion inputs has close similarities among unrelated species, there is evidence of differences between species both in their susceptibility to motion sickness and also in the pharmacological effects of drugs that prevent motion sickness.

Neuropharmacology of Motion Sickness

All drugs that are active in motion sickness have a central rather than peripheral mode of action. Both the centrally acting anticholinergic hyoscine and the central adrenergic drugs amphetamine and ephedrine are effective. By contrast, the phenothiazine drugs, which are thought to prevent vomiting through their blocking action on dopamine D_2 receptors in the CTZ, are in general of little effect in motion sickness. This tends to suggest that an important step, specific to the development of motion sickness, involves muscarinic cholinergic receptors.

Though the precise site of action of anti-motion sickness drugs is not known, it may be significant that a reciprocal action between acetylcholine and noradrenaline is found in many cells of the vestibular nuclei. In cats, an excitatory effect from acetylcholine was found in 80% of cells sampled and an inhibitory effect from noradrenaline and amphetamine was found in 75% of cells (Kirsten and Sharma 1976). Furthermore, both hyoscine and diphenhydramine (an antihistaminic effective in motion sickness), have been shown to suppress spontaneous and motion-induced activity in the cat vestibular nucleus. It is thought that the role of histamine is to increase cholinergic activity through an action on H1 receptors located on certain cholinergic neurons (Nowak et al. 1977). Blockade of these H1 receptors by some antihistamines may explain their prophylactic effect in motion sickness.

Further evidence of the significance of cholinergic pathways in motion sickness may be provided by the observation that centrally acting cholinesterase inhibitors, such as physostigmine, produce a pattern of physiological response that resembles that of motion sickness (Davis and Davis 1980). This response can be antagonised by many of the drugs that are effective in motion sickness (Janowski et al. 1984).

The vestibular nerves also carry efferent fibres to the hair cells of the otolithic maculae and semi-circular canals. Doubt remains about their function (reviewed by Precht 1974), but it is thought that stimulation of these efferents may have a widespread excitatory effect on the vestibular hair cells (Goldberg and Fernandez 1980). Inhibition of vestibular efferents by drugs might therefore play a part in preventing motion sickness by reducing incoming signals from the vestibular end organ. However, the vestibular efferent fibres are thought to terminate as nicotinic cholinergic synapses and would therefore not be blocked by hyoscine.

Treatment of Motion Sickness

Preventive Measures

Since the generation of sensory conflict underlies all motion environments that give rise to motion sickness, practical measures that reduce conflict are likely to reduce motion sickness incidence.

Visual-vestibular conflict can be minimised if the subject has the widest possible view of an earth-stable visual reference. The passenger aboard ship is less liable to be sea sick if he remains on deck, somewhere amidships to minimise vertical motion, and views the horizon. In a car or coach, sufferers should be in a position to see the road directly ahead (or behind), since the movements of this visual scene will accord with the changes in direction of the vehicle. Conflict between semi-circular canals and otoliths is generated in vehicles cornering at speed and can also result from head movements made in a rotating environment. Because radial accelerations in cornering vehicles are proportional to the square of forward velocity of the vehicle, comparatively small reductions in speed on twisting roads may significantly reduce the incidence of motion sickness in passengers. In passenger aircraft, turns are generally of large radius of curvature and correspondingly low angles of bank and the consequent mismatch is small. In military aircraft, this is often not the case. There is some evidence that the incidence of air sickness in paratroopers is reduced by the use of restraints to limit head movement, though no significant beneficial effect was found in turbulent flight conditions which impose a mainly vertical oscillatory motion on the aircraft (Johnson and Mayne 1953), and a similar study by Keist et al. (1956) failed to show any benefit from head restraint. While head movements in a rotating environment are known to precipitate motion sickness, there is no clear experimental evidence that they potentiate the nauseogenic effects of low-frequency linear oscillation. In such conditions, notably on larger ships, body orientation with respect to the direction of alternating acceleration has a significant effect. In a study using an aircraft to generate this stimulus by means of roller-coaster type manoeuvres, susceptibility to motion sickness was less in subjects sitting upright, with their head and trunk aligned with the changing force vector, than in those sitting with the head and trunk flexed forward to the horizontal (von Baumgarten et al. 1980 a). A similar oscillating acceleration can be generated in road vehicles by alternately accelerating and braking the vehicle. This stimulus is also nauseogenic for the occupants, more so for sitting passengers than for those lying down (Vogel et al. 1982).

Drug Prophylaxis

A large number of drugs have been tested for their possible effect in motion sickness prophylaxis (Brand and Perry 1966; Wood and Graybiel 1968). Those that are of benefit fall into several different pharmacological groups. Their therapeutic effectiveness in motion sickness does not, however, parallel their in vitro potency within any pharmacological group. For example, despite equivalent anti-cholinergic activity, hyoscine is considerably more effective than atropine. This difference is attributed to the greater central action of hyoscine. Some antihistamines, such as dimenhydrinate, cyclizine and cinnarizine are effective in motion sickness, but many are not. The phenothiazine derivatives also possess antihistaminic properties, but of this group only promethazine is particularly effective.

Therapeutic trials of prophylactic drugs present a number of problems. No suitable animal model is available for the screening of new drugs. Rabbits and cats have a low motion sickness susceptibility. That of dogs is similar to man but drugs known to be effective in man, such as hyoscine, diphenhydramine and promethazine, are ineffective in preventing swing-induced sickness in dogs (Chinn and Plotnikoff 1953), though meclozine has been shown to be effective (Turnbull and Chinn 1960).

Drug trials carried out at sea or in troop-carrying aircraft have often yielded conflicting results, possibly as a result of an inability to control the stimulus conditions. Laboratory stimuli enable a more controlled stimulus dose, but fewer subjects can be tested at one time and stimulus levels may be high for short periods compared with the longer duration lower intensity motion environment of a ship. A further source of difference in comparative trials between drugs may be related to their time course of action. Thus, the beneficial effect of hyoscine is probably at its maximum 1–2 h after oral administration and has fallen to low levels by 6 h (Brand 1969). By contrast, promethazine does not achieve maximum blood levels until 3–5 h after ingestion, but continues to have a therapeutic effect for more than 24 h.

The dose/response relationship for prophylactic drugs has also received insufficient attention and is known with any accuracy only for hyoscine. An increase in drug dosage does not necessarily yield a corresponding increase in prophylactic effect and in some drugs a higher than normal dose has resulted in a reduced effectiveness (Wood 1965).

The drug L-hyoscine hydrobromide (hyoscine, scopolamine) emerged as a valuable prophylactic drug following extensive research during World War II into the problems of motion sickness in troops transported in aircraft, ships and landing-craft. It remains one of the most effective drugs for short duration exposures to provocative motion. Doses in excess of 0.6 mg are, however, very liable to lead to drowsiness and there is experimental evidence that hyoscine impairs short-term memory (Crow 1979). Such side effects may be acceptable, even desirable, in passengers, but they severely limit the usefulness of the drug in, for example, aircrew under training. Hyoscine can be absorbed transdermally, and in order to extend the duration of action, a controlled-release patch has been developed to deliver 1.2 mg on application and 0.01 mg hourly thereafter (Shaw and Chandrasekaran 1978). There is good evidence of its sustained effectiveness but, perhaps owing to variable absorption rates, there is an increased risk of blurred vision after more than 24 h of use (Homick et al. 1983).

Amphetamine was originally used in conjunction with hyoscine to counteract its soporific effects, but has been shown to have a prophylactic effect on motion sickness when used alone (Hill 1937), and to be synergistic both with hyoscine and also promethazine (Graybiel et al. 1975). Ephedrine 25 mg has comparable effects to amphetamine 10 mg and is to be preferred for repeated use as it is less liable to lead to habituation.

The chance observation in 1949 that a patient who was receiving dimenhydrinate for urticaria was also relieved of sea sickness has led to the discovery of a number of antihistamines that are clinically useful in motion sickness. Following oral administration, these drugs are generally slower than hyoscine in reaching their peak efficacy but have a longer duration of action. Their relative effectiveness is not clearly established but they differ in the incidence of sedative and anticholinergic side effects.

Promethazine 25 mg with either amphetamine 10 mg ephedrine 25 mg has been shown to be superior to hyoscine or other antihistamines used alone (Wood

1970). There is no experimental work to indicate whether centrally-acting sympathomimetics exhibit a similar synergism with other antihistamines.

Cinnarizine has recently achieved popularity among the sailing fraternity as an effective drug against sea sickness, and it was also widely used in naval operations during the Falklands War. in sea-going trials, cinnarizine has been shown to be as effective as hyoscine and to produce fewer side effects (Hargreaves 1982), though the 6–8 hourly administration of drugs used in the trial may not, in the case of hyoscine, have been sufficiently frequent to maintain an adequate level of protection. Laboratory trials using a cross-coupled provocative stimulus 2 h after drug administration showed cinnarizine 15 mg to be somewhat less effective than hyoscine 0.6 mg (Wood and Graybiel 1968; Stott et al. 1984).

Metoclopramide 20 mg has also been shown to have a prophylactic effect in airsickness induced by "roller-coaster" flight manoeuvres (von Baumgarten et al. 1980 b). This drug has been used in the prophylaxis of space sickness though no clear evidence for its value in this environment has been obtained (Oman et al. 1984).

Desensitisation Treatment

For most susceptible subjects, whose exposure to motion sickness-inducing stimuli is infrequent, prophylactic drugs offer the only useful treatment. When exposure to provocative stimuli is more frequent, as for example in professional aircrew, spontaneous adaptation occurs during training and an initially high incidence of motion sickness decreases with time (Tucker et al. 1965). Those whose rate of spontaneous adaptation is slow can be helped by the use of desensitisation treatment. A number of regimes have been described, but all make use of graded exposure to laboratory-based stimuli to induce a level of malaise that falls well short of vomiting. Adequate desensitisation requires a period of up to 4 weeks. In some schemes, use is made of sensory feedback of such autonomic variables as heart rate, finger plethysmography and galvanic skin resistance (Cowings and Toscano 1982).

This technique appears to increase the rate of adaptation to a cross-coupled stimulus. In the Royal Air Force, an initial desensitisation to laboratory stimuli is followed by a period of dual flying under the control of a medical officer pilot before air crew are returned to their flying training. By this means, control of motion sickness has been achieved in 84% of subjects (Stott and Bagshaw 1984). Desensitisation treatment is both expensive and time consuming, but is justified if an even more expensive training would otherwise be wasted.

A similar need arises in the costly environment of space, where space sickness may decrease the efficiency of astronauts in the first few days of a mission. Prespaceflight experience of the zero-*g* environment is obtained for short periods by means of parabolic flight manoeuvres in a KC 135 aircraft. Despite showing tolerance to the nauseogenic effects of these manoeuvres, astronauts do not achieve complete pre-flight desensitisation, and space sickness remains a problem to almost half the astronauts during the first few days of space flight (Oman 1982).

References

Bard P, Woolsey CW, Snider RS, Mountcastle VB, Bromley RB (1947) Delimination of central nervous mechanisms involved in motion sickness. Fed Proc 6:72
Benson AJ (1973) Use of nystagmography in the study of aircrew with spatial disorientation. In: The use of nystagmography in aviation medicine. AGARD-CP-123: A4 1-13 AGARD/NATO, Neuilly-sur-Seine
Benson AJ (1977) Possible mechanisms of motion and space sickness. Proceedings of the European symposium on life science research in space. ESA SP-130:101–108
Brand JJ (1969) The time course of action of l-hyoscine after intramuscular injection. Br J Pharmacol 35:202–208
Brand JJ, Perry WLM (1966) Drugs used in motion sickness. Pharmacol Rev 18:895–924
Brooks M (1939) The etiology of seasickness. Med Rec NY 150:23–26
Chinn NI, Plotnikoff BP (1963) Evaluation of various techniques for screening anti-motion sickness drugs. J Appl Physiol 5:392–394
Claremont CA (1931) The psychology of seasickness. Psyche (Stuttg) 11:86–90
Cowings P, Toscano WB (1982) The relationship of motion sickness susceptibility to learned autonomic control for symptom suppression. Aviat Space Environ Med 53:570–575
Crow TJ (1979) Action of hyoscine on verbal learning in man: Evidence for a cholinergic link in the transition from primary to secondary memory. In: Brazier MAB (ed) Brain mechanisms in memory and learning: from the single neuron to man. Raven, New York
Davis BM, Davis KL (1980) Cholinergic mechanisms and anterior pituitary hormone secretion. Biol Psychiatry 15:303–310
Dobie TG (1974) Airsickness in aircrew. AGARDograph AG-177, AGARD/NATO, Neuilly-sur-Seine
Doig RK, Wolf S, Wolff MS (1953) Study of gastric function in a 'decorticate' man with a gastric fistula. Gastroenterology 23:40–44
Dufosse M, Ito M, Miyashita Y (1977) Functional localisation in the rabbit's cerebellar flocculus determined in relationship with eye movements. Neurosci Lett 5:273–277
Eversmann T, Gottsmann M, Uhlich E, Ulbrecht G, von Werder K, Scriba PC (1977) Increased secretion of growth hormone, prolactin, antidiuretic hormone and cortisol induced by the stress of motion sickness. Aviat Space Environ Med 49:53–57
Fernandez C, Goldberg JM (1971) Physiology of peripheral neurons innervating semicircular canals of the squirrel monkey. II Response to sinusoidal stimulation and dynamics of peripheral vestibular system. J Neurophysiol 34:661–675
Goldberg JM, Fernandez C (1980) Efferent vestibular system in the squirrel monkey. Anatomical location and influence on efferent activity. J Neurophysiol 43:986–1025
Graybiel A (1970) Susceptibility to acute motion sickness in blind persons. Aerospace Med 44:593–608
Graybiel A, Wood CD, Miller EF, Cramer DB (1968) Diagnostic criteria for grading the severity of acute motion sickness. Aerospace Med 39:453–455
Graybiel A, Miller EF, Homick JL (1974) Experiment no 131. Human vestibular function. In: Proceedings of the skylab life sciences symposium. JSC-09275 NASA TM X-58154. NASA, Lyndon B Johnson Space Centre, Houston
Graybiel A, Wood CD, Knepton J, Hoche JP, Perkins GF (1975) Human assay of anti-motion sickness drugs. Aviat Space Environ Med 46:1107–1118
Guedry FE (1970) Conflicting sensory cues as a factor in motion sickness. In: 4th Symposium on the role of the vestibular organs in space exploration. NASA SP-187: 45–52. Office of Technology Utilisation, NASA, Washington, DC
Hargreaves J (1982) The prophylaxis of seasickness, a comparison of cinnarizine with hyoscine. Practitioner 226:160
Hill H (1937) Benzedrine in seasickness. Br Med J 2:1109–1112
Hill T (1984) The relationship between motion sickness susceptibility and the adaptation to post-rotary stimulation. Thesis, Coventry Polytechnic, UK
Homick JL (1979) Space motion sickness. Acta Astronautica 6:1259–1272

Homick JL, Kohl RL, Reschke MF, Degioanni J, Cintron-Trevino NM (1983) Transdermal scopolamine in the prevention of motion sickness: Evaluation of the time course of efficacy. Aviat Space Environ Med 54:994–1000
Irwin JA (1881) The pathology of seasickness. Lancet 2:907–909
Ito M, Miyashita Y (1975) The effects of chronic destruction of the inferior olive upon visual modification of the horizontal vestibulo-ocular reflex of rabbits. Proc Jpn Acad 51:716–720
James W (1882) The sense of dizziness in deaf-mutes. Am J Otol 4:239–254
Janowski DS, Risch SC, Ziegler M, Kennedy B, Huey L (1984) A cholinomimetic model of motion sickness and space adaptation syndrome. Aviat Space Environ Med 50:1046–1051
Johnson WH, Mayne JW (1953) Stimulus required to produce motion sickness. Restriction of head movement as a preventive of airsickness – field studies on airborne troops. J Aviat Med 24:400–411
Keist BF, Shelley WT, Byers JM, Chinn HI (1956) Effect of head immobilisation on incidence of airsickness. J Appl Physiol 8:369–370
Kirsten EB, Sharma JN (1976) Characteristics and response differences to iontophoretically applied norepinephrine, d-amphetamine and acetylcholine on neurons in the medial and lateral vestibular nuclei of the cat. Brain Res 112:77–90
Kottenhoff H, Lindahl LE (1960) Laboratory studies on the psychology of motion sickness. Acta Psychol 17:89–112
Lawther A, Griffin MJ (1981) Motion sickness in sea-going passenger vessels: an interim report. Human Factors Research Unit, ISVR Southampton
Lisberger SG, Fuchs AF (1978) role of primate flocculus during rapid behavioural modification of vestibulo-ocular reflex. I. Purkinje cell activity during visually guided horizontal smooth pursuit eye movements and passive head rotation. J Neurophysiol 41:733–763
McCauley ME, Jackson WR, Wylie CD, O'Hanlon JF, Mackie RR (1976) Motion sickness incidence: exploratory studies of habituation, pitch and roll, and the refinement of a mathematical model. Technical Report 1733-2, Human Factors Research Inc, Santa Barbara Research Park, Goleta
McClure JA, Fregley AR, Molina E, Graybiel A (1972) Response from arousal and thermal sweat areas during motion sickness. Aerospace Med 43:176–179
Minor JL (1896) Seasickness: its cause and relief. NY Med J 64:522–523
Money KE, Cheung BS (1983) Another function of the inner ear: facilitation of the emetic response to poisons. Aviat Space Environ Med 54:208–211
Money KE, Wood JD (1970) Neural mechanisms underlying the symptomatology of motion sickness. In: 4th Symposium on the role of the vestibular organs in space exploration. NASA SP-187: 35–44. Office of Technology Utilisation, NASA, Washington, DC
Moore HJ, Lentz JM, Geudry FE (1977) Nauseogenic visual-vestibular interation in a visual search task. Report NAMRL-1234. Naval Aerospace Medical Research Lab, Pensacola
Nowak JZ, Pilc A, Lebrecht U, Maslinski C (1977) Does histamine interact with cholinergic neurons in its cataleptogenic action in the rat? Neuropharmacology 16:841–847
O'Hanlon JF, McCauley ME (1974) Motion sickness incidence as a function of the frequency of vertical sinusoidal motion. Aerospace Med 45:366–369
Oman CM (1982) Space motion sickness and vestibular experiments in spacelab. Paper No 820833, Soc Automotive Eng, 12th Intersoc conference on environmental systems
Oman CM, Lichtenberg BK, Money KE (1984) Space motion sickness monitoring experiment: Spacelab 1. In: Motion sickness: mechanisms, prediction, prevention and treatment. AGARD-CP-372: 35, 1-21. AGARD/NATO, Neuilly-sur-Seine
Precht W (1974) Physiological aspects of the efferent vestibular system. In: Kornhuber HH (ed) Vestibular system, part 2. Psychophysics, applied aspects and general interpretation. Springer, Berlin Heidelberg New York, pp 221–236 (Handbook of sensory physiology, vol 4)
Reason JT (1968) Relation between motion sickness susceptibility, the spiral after-effect and loudness estimation. Br J Psychol 59:385–393
Reason JT (1970) Motion sickness: a special case of sensory re-arrangement. Adv Sci 26:386–393
Reason JT (1978) Motion sickness adaptation: a neural mismatch model. J R Soc Med 71:819–829

Reason JT, Brand JJ (1975) Motion sickness. Academic, London, pp 174–181
Reason JT, Graybiel A (1969) Adaptation to coriolis accelerations: its transfer to the opposite direction of rotation as a function of interventing activity at zero velocity. NAMI-1086, NASA Order R-93, Naval Aerospace Medical Institute, Pensacola
Shaw JE, Chandrasekaran KR (1978) Controlled topical delivery of drugs for systemic action. Drug Metab Rev 8:223–233
Sinha R (1968) Effect of vestibular coriolis reaction on respiration and blood flow changes in man. Aerospace Med 39:837–844
Stott JRR, Bagshaw M (1984) The current status of the RAF programme of desensitisation for motion sick aircrew. In: Motion sickness: mechanisms, prediction, prevention and treatment. AGARD-CP-372:40, 1–9. AGARD/NATO, Neuilly-sur-Seine
Stott JRR, Hubble MP, Spencer MB (1984) A double blind comparative trial of powdered ginger root, hyoscine hydrobroide and cinnarizine in the prophylaxis of motion sickness induced by cross coupled stimulation. In: Motion sickness: mechanisms, prediction, prevention and treatment. AGARD-CP-372:39, 1–6. AGARD/NATO, Neuilly-sur-Seine
Triesman M (1977) Motion sickness: an evolutionary hypothesis. Science 197:493–495
Tucker GJ, Hand DJ, Godbey AL, Reinhardt RF (1965) Airsickness in student aviators. NSAM-939. Naval School of Aviation Medicine, Pensacola
Turnbull R, Chinn HL (1960) The effect of certain drugs on the incidence of seasickness. Clin Pharmacol Ther 1:280–283
Vogel H, Kohlhaas R, von Baumgarten RJ (1982) Dependence of motion sickness in automobiles on the direction of linear acceleration. Eur J Appl Physiol 48:399–405
Von Baumgarten RJ, Baldrighi G, Vogel H, Thumler R (1980a) Physiological responses to hyper- and hypo-gravity during roller coaster flight. Aviat Space Environ Med 51:145–154
Von Baumgarten RJ, Thumler R, Vogel H (1980b) Experimentelle Untersuchungen über die Wirksamkeit von Metoclopramid bei Kinetose. Therapiewoche 30:5974–5981
Waespe W, Henn V (1977) Neuronal activity in the vestibular nuclei of the alert monkey during vestibular and optokinetic stimulation. Exp Brain Res 27:523–538
Waespe W, Henn V (1978) Conflicting visual-vestibular stimulation and vestibular nucleus activity in alert monkeys. Exp Brain Res 33:203–211
Wang SC, Borison HL (1952) A new concept of the organisation of the central emetic mechanism: recent studies on the sites of action of apomorphine, copper sulphate and cardiac glycosides. Gastroenterology 22:1–12
Wang SC, Chinn HI (1954) Experimental motion sickness in dogs. Functional importance of chemoceptive emetic trigger zone. Am J Physiol 178:111–116
Wang SC, Chinn HI (1956) Experimental motion sickness in dogs. Importance of labyrinth and vestibular cerebellum. Am J Physiol 185:617–623
Wang SC, Chinn HI, Renzi AA (1957) Experimental motion sickness in dogs: role of abdominal visceral afferents. Report No 57-112, School of Aviation Medicine, Randolph AFB, Texas
Wilding JM, Meddis R (1972) Personality correlates of motion sickness. Br J Psychol 63:619–620
Wood CD (1965) Anti-motion sickness drugs for aerospace. In: 1st Symposium on the role of the vestibular organs in the exploration of space. NASA SP-77:365–371. Scientific and Technical Information Division, NASA, Washington, DC
Wood CD (1970) Anti-motion sickness therapy. In: 5th Symposium on the role of the vestibular organs in space exploration. NASA SP-314:109–114. Scientific and Technical Information Office, NASA, Washington, DC
Wood CD, Graybiel A (1968) Evaluation of sixteen anti-motion sickness drugs under controlled laboratory conditions. NAMI-982, Naval Aerospace Medical Institute, Pensacola
Young LR, Oman CM, Watt DGD, Money KE, Lichtenberg BK (1984) Spatial orientation in weightlessness and re-adaptation to earth's gravity. Science 225:205–208

Mechanisms and Treatment of Post-Operative Nausea and Vomiting

R. F. Cookson [1]

Introduction

This review of post-operative nausea and vomiting discusses the principal aetiological features together with clinical research findings relevant to prophylaxis and treatment of the problem. However, most aetiological factors interact with each other so that, for example, detailed changes may be made in the anaesthetic to allow for changes in patient age or type of operation. Most studies have ignored such variations resulting from, for example, changes in premedication or, when more carefully defined patient groups are investigated, have used insufficient numbers of patients to reach significant conclusions.

Table 1. Overall incidence of post-operative emetic sequelae

Source		Population	Percent
Waters	(1936)	10,638	40.6
Dent et al.	(1955)	2,000	27.2
Burtels and Peckett	(1957)	1,702	32.0
Bonica et al.	(1958)	1,561	30.5
Bellville et al.	(1960)	1,569	17.0
Adriani et al.	(1961)	2,230	23.0

The contribution of clinical research to this area is marred by a lack of clear thinking. Progress will continue to be limited until one is able to ask a simple question and get an unequivocal answer. Too frequently confusion and contradiction abound because of inappropriate pooling of data from situations which are inherently not comparable (see Table 1).

When the older anaesthetics ether and cyclopropane were in frequent use, post-operative nausea and vomiting was a significant problem. More recently, these agents have been replaced by anaesthetics such as halothane, which may have significantly less emetic potential, or by the careful use of analgesic supplemented nitrous oxide techniques, reducing the incidence of nausea and vomiting. There have been no recent studies with large numbers of patients to assess the impact of these changes in technique. The differences in Table 1 reflect not only changes in the anaesthetic, but also the use of different criteria (e.g. definitions of nausea and vomiting, length of follow-up, patient population, and closeness of observation of the patient during the postoperative period). There has been

1 Medical Department, Janssen Pharmaceutical Limited, Grove, Wantage, Oxfordshire OX12 0DQ, UK

Nausea and Vomiting: Mechanisms and Treatment
Ed. by C. J. Davis, G. V. Lake-Bakaar and D. G. Grahame-Smith

little attempt by most authors to define or even differentiate between nausea, retching and vomiting, some or all of which are occasionally grouped together as "post-operative sickness".

Historical comparisons suffer from changes in population and premedication between published studies. Although aetiological factors are frequently interdependent, most studies pay insufficient attention to this source of confusion. The incidence of nausea, retching and/or vomiting will vary as a result of the interplay between various factors.

This review considers three main groups of factors: the influence of premedication and anaesthetic technique directly (Sect. 3.1); the patient and type of operation (Sect. 3.2); the post-operative period (Sect. 3.3). In addition to these factors, it is important not to forget when assessing the literature, the fourth, often unstated influence of the investigator. Results are easily distorted by arbitrary decisions on criteria for nausea, retching and vomiting, the definition of these symptoms, and the scoring method employed to rate them. The assessment of nausea is particularly problematical because the patient must be questioned and the manner of questioning and the frequency of observation of the patient will have some bearing on his subjective response. Much post-operative morbidity occurs in the first hour, and many studies have reported data based on observations for the first 2 or 3 h post-operatively. It is clear, however, from more extended studies that emetic sequelae continue for at least 24 h post-operatively.

Much of the literature is confused and contradictory, often as a result of pooling data. The type of operation will bias, for example, the ratio of men to women in the population and age-related conclusions may be distorted by modification of the anaesthetic technique, particularly for the young and the old. A particular anaesthetic may be selected for a specific operation and it is then difficult to compare the incidence of emetic problems as a function of changes in the site of operation. In spite of these confusing factors, certain groups seem particularly liable to emetic problems, for example female obese patients, or those with a history of motion sickness or post-operative sickness from other anaesthetics (Purkis 1964).

Nausea and vomiting may be expected in some series to occur in up to half the patients and some regard vomiting as the worst feature of the post-operative period (Riding 1975; Cronin et al. 1973). Mild vomiting, although distressing to the patient, does not usually lead to electrolytic disturbances. Severe vomiting on the other hand may produce marked changes in electrolytes and result in dehydration, particularly in children. Fortunately, this complication is very uncommon after anaesthesia (Palazzo and Strunin 1984a; Smith and Manford 1974). A detailed discussion of the electrolyte changes following severe vomiting is given by Clarke (1984).

Although patients are frequently fasted before anaesthesia and surgery, Bodman and colleagues (1960) have pointed out there are still significant solid gastrointestinal contents available in some patients for quite copious vomiting to occur in the post-operative period.

Adriani and his co-workers (1961) pointed out that only 3.5% of patients had severe persistent vomiting that would require treatment. By far the largest percentage of vomiting occurred on emergence from anaesthesia and did not recur thereafter. In addition, the patients did not remember these events and therefore they were not important. In contrast, Clarke (1984) has emphasised the high risk associated with postoperative vomiting if laryngeal reflexes are impaired and the trachea is not adequately protected.

Aspiration of vomitus into the trachea is the most worrying problem associated with post-operative nausea and vomiting, particularly as this may occur in the presence of depressed consciousness or, for example, when the jaw is wired after oral surgery. Another major concern following delicate surgery such as that of the eye or plastic surgery, is wound disruption due to the physical stresses associated with retching and vomiting. The hazard of vitreous loss following intraocular operations is well recognised. More recently the specific problems associated with plastic surgery have been reviewed (Stein 1982). The high intra-abdominal pressure associated with retching and vomiting results in renal hypertension and an associated significant increase in bleeding beneath skin flaps.

Mechanisms

The seminal paper by Borison and Wang (1953) still dominates current thinking on the mechanisms of nausea and vomiting. The mechanisms of anaesthetic-induced vomiting have been little studied, although it seems that both general anaesthetics and the opiates used in pre-medication and perioperatively may interact with both the chemoreceptor trigger zone (CTZ) and the vomiting centre. At some doses these agents appear to affect the recently postulated "anti-emetic centre" (Costello and Borison 1977). The opiate interaction with the vomiting centre may be mediated by vestibular sensitisation, but as the main reason for using opiates during an operation or in premedication is to reduce the amount of pain postoperatively, this effect is confused by the emetic effect of pain on the vomiting centre.

Other influences on the vomiting centre are physical stimulation of the oropharynx, particularly by inexperienced anaesthetists and psychic disturbances due to patient awareness (e.g. during surgery under local anaesthesia or after short interventions in out-patient clinics). In addition, peroperative and postoperative hypotension may also elicit vomiting. The ubiquitous anaesthetic nitrous oxide is suggested to have a direct effect on the vomiting centre, although it may be that its more significant action is by way of gastro-intestinal distension caused either by careless manual ventilation or by direct transfer of the gas into the gastrointestinal tract (Palazzo and Strunin 1984a).

It is clear that anaesthetic and surgical factors interact both with the vomiting centre and CTZ and with each other. Nimmo (1984) has pointed out that the opioids are the main cause of delay in gastric emptying in the peri-operative period, and that this is presumably associated with an increase in nausea and vomiting. Volatile anaesthetics either directly or by gastric distension may stimulate stomach vagal afferents, resulting in a high incidence of vomiting post-operatively (Bellville 1961). Although the use of gastric tubes reduces this incidence significantly, more systematic work is required in this area (Palazzo and Strunin 1984a).

Aetiology

Anaesthesia and Pre-Medication

A variety of interdependent factors affect the anaesthetic contribution to post-operative nausea and vomiting. The difficulties in assessing the role of each component are compounded by the lack of careful controlled studies in this area, and the need to vary the details of the anaesthetic technique for particular operative circumstances. In addition to the factors discussed below, Palazzo and Strunin (1984a) have pointed to the little investigated effects of the use of oral airways, and the occurrence of perioperative hypoxia or carbon dioxide retention on the incidence of post-operative nausea and vomiting.

Spinal and Regional Anaesthesia

Dent et al. (1955) reported an incidence of vomiting of 11.1% following spinal anaesthesia and only 4.3% after regional block. As local anaesthesia removes most of the emetic stimuli associated with general anaesthetics, it is not surprising that in careful hands these techniques are associated with a marked reduction in the incidence of vomiting post-operatively, particularly as the use of 100% oxygen reduces the risk of hypoxia at the vomiting centre (Ratra et al. 1972).

Other authors (Crocker and Vandam 1959; Bonica et al. 1958) have reported a higher incidence of nausea and vomiting following spinal anaesthesia and have indicated that the incidence varies depending on the site of operation and the degree of hypotension and pain experienced by the patient. Datta et al. (1982) showed that if hypotension, which is a frequent complication of spinal anaesthetics, is treated with i.v. ephedrine, then the percentage of nausea and vomiting is kept to 10%. In contrast, if the treatment of the hypotension is delayed, then the incidence of nausea and vomiting increases to 66%. Although there are advantages in being able to perform anaesthesia under 100% oxygen, there has been little work on the possibility of the patient suffering from nausea and vomiting as a result of psychological and visual stimulation. If a regional anaesthetic requires at some stage the addition of general anaesthesia, then the percentage of vomiting increases to more than that expected from the separate contributions of the spinal and general anaesthetic (Bonica et al. 1958).

Nitrous Oxide

Parkhouse and colleagues (1960) indicated that the nausea experienced by volunteers exposed to nitrous oxide was related to the concentration of inspired gas. As the percentage of nitrous oxide increased from 20 to 30 to 40 percent the incidence of nausea increased. Bodman et al. (1960) demonstrated that each minute of nitrous oxide anaesthesia was important as an influence on the incidence of post-operative vomiting.

Table 2. Incidence (%) of emetic sequelae in the first 6 h post-operatively as a function of pre-medication and duration of anaesthesia for D and C (Morrison et al. 1968)

Pre-medication	Duration of anaesthetic	
	2–4 min	8–10 min
Non-opiate (includes anti-emetic)	14	28
Opiates of low emetic potential	28	43
Opiates of high emetic potential	50	58

Nitrous oxide, which is the most ubiquitous of the anaesthetics, is rarely used alone and the amount of nausea and vomiting post-operatively depends on the other agents used with the gas. Dundee et al. (1965) evaluating techniques for patients presenting for minor gynaecological surgery, demonstrated a significant difference in the emetic sequelae between nitrous oxide supplemented with intermittent thiopentone and the use of nitrous oxide with intermittent propanidid. Morrison et al. (1968), comparing patients presenting for dilatation and curettage (D and C) and curettage, demonstrated significant differences in the percentage of emetic sequelae depending on the premedication used (see Table 2). In each group, the incidence of nausea and vomiting increased with the duration of anaesthesia, and in each case the incidence was higher following D and C than following curettage alone.

In longer operations, generally done under the influence of muscle relaxants, the role of nitrous oxide is further confused by the effect of the relaxant and reversal agent that may be used, as well by supplementary doses of opioids during the anaesthetic. Kvisselgaard (1958) found no effect of duration of anaesthesia on the incidence of nausea and vomiting when a careful opioid nitrous oxide relaxant technique was used. Janhunen and Tammisto (1972) reported an incidence of approximately 60% of emetic problems following various nitrous oxide relaxant techniques; whereas, Riding (1963) anecdotally reported a variation in incidence ranging from 11% for pneumonectomy in males to 80% for major abdominal gynaecological operations.

Inhalational Agents

Bellville and his co-workers (1960) reported an increased incidence of emetic sequelae after cyclopropane anaesthesia in women as the duration of the anaesthetic increased (see Table 3). Palazzo and Strunin (1984a) point out that longer operations not only increase the exposure of the patient to potential emetic agents, but they also reduce the effect of any ameliorative drug given in the premedication. In addition, there is a significant increase in the time since the last meal and there may well be an influence on the patient akin to that of starvation nausea reported in volunteers. Bellville (1961) postulated a dose-response rela-

Table 3. The influence of duration of anaesthesia in women receiving cyclopropane (Bellville et al. 1960)

Duration (min)	Incidence (%)
30– 90	17.5
90–150	28.4
150–210	46.4

tionship for inhalational agents as in his view any step which reduced the amount of anaesthetic given to the patient was associated with a reduction in post-operative vomiting. None of the features he listed have been adequately investigated, although there is much anecdotal support for the various factors suggested. According to Bellville (1961), the factors minimising exposure to inhalation anaesthetics and reducing vomiting are:

1. Short duration of anaesthesia
2. Endotracheal intubation
3. The use of muscle relaxants
4. Experienced anaesthetists
5. Light anaesthesia
6. Extra-abdominal procedures
7. Thin patients

Early work on different inhalational agents lacked strict methodology so that results can rarely be compared as different criteria have been used to assess differing populations. Relatively few comparative studies have been carried out and there are none on the newer agents such as enflurane and isoflurane. The older agents, cyclopropane and ether, have been associated with an incidence of nausea and vomiting ranging from 8%–85%, contrasting unfavourably with a somewhat lower incidence reported for the most recent studies for halothane.

It is interesting to compare the data on halothane from the studies by Heal (1965) and Dundee et al. (1965). Heal reported an incidence of nausea following halothane anaesthesia of only 6%, but 26% of his patients vomited more than once after the anaesthetic. In contrast, Dundee's incidence of nausea was higher (23%), but only 10% of his patients vomited in the post-operative period. It may be that the differences between these two studies is entirely a result of the different premedication used. Atropine was used in Dundee's series and atropine with pethidine by Heal (see Table 4).

The influence of pre-medication (see Sect. 3.1.5) is further illustrated by Dundee et al. (1965) who compared halothane and cyclopropane over the first post-operative hour and over the succeeding 5 h (see Table 5). The influence of cyclopropane on postoperative emesis independent of premedication is clearly illustrated when compared with halothane anaesthesia. Patients, however, continue to vomit after the 6th hour post-operatively and although Purkis (1964) was able to show statistically significant differences between halothane and cyclopropane when data to 6 h were collected, these differences were somewhat reduced when the full 24 h data were compared. Interpretation of data on halothane is even fur-

Table 4. Variation of % incidence of emetic sequelae with inhalational agent

Source	N/R/V	Ether	Cyclo-propane	Halothane	Trilene
Howat (1960)	N/V	–	–	41.7	65.5
Waters (1936)	N/V	56.5	39.0	–	–
Dent et al. (1955)	R/V	38.5	34.6	–	–
Burtels et al. (1957, 1960)	V	33.4	23.9	28.0	26.1
Purkis (1964)	V	33.9	32.3	26.8	–
Riding (1963)	E	35.0	–	5.0	7.5
Coppolino and Wallace (1962)	E	34.3	8.0	3.4	–
Haumann and Foster (1963)	V	–	–	4.5	37.1
Dundee et al. (1965)	N/V	–	85.0	33.3	–
Dundee et al. (1965)	V	–	–	10.0	–
Heal (1965)	N/V	–	–	40.0	–
Heal (1965)	V	–	–	34.0	–

N, nausea; R, retching; V, vomiting; E, emergence vomiting

Table 5. Emetic sequelae following halothane or cyclopropane (Dundee et al. 1965)

Anaesthetic	Pre-medication	Nausea (%)		Vomiting (%)	
		1st hour	1–6 h	1st hour	1–6 h
Halothane	Atropine	21.7	6.7	6.7	3.3
	Pethidine + atropine	16.7	10.0	10.0	6.7
	Morphine + atropine	13.3	10.0	30.0	26.7
Cyclopropane	Atropine	16.7	6.7	67.0	6.7
	Pethidine + atropine	5.0	30.0	85.0	10.0
	Morphine + atropine	25.0	10.0	65.0	40.0

ther confused by the observation (Haumann and Foster 1963) that at sub-anaesthetic doses, halothane has an anti-emetic effect.

Although volatile anaesthetics contribute significantly to the increased incidence of nausea and vomiting post-operatively, Palazzo and Strunin (1984a) have pointed out that avoiding these does not necessarily decrease emetic sequelae as there is a tendency to replace inhalational agents with potent narcotics. Janhunen and Tammisto (1972) scoring for nausea, retching and vomiting under standard conditions, were unable to find significant differences between techniques of anaesthesia involving relaxants halothane and nitrous oxide, relaxants pethidine and nitrous oxide and a nitrous oxide relaxant technique without supplementation, although as Palazzo and Strunin (1984a) point out, the dose of pethidine used may have exerted an anti-emetic effect (see Sect. 3.1.5).

Intravenous Agents

Clarke (1984) has reported data on various intravenous agents given intermittently under standard conditions for minor gynaecological procedures as a sup-

Table 6. Incidence of nausea and vomiting (%) in first 6 post-operative hours (N_2O/O_2 anaesthesia with intermittent supplementation; Clarke 1984

	Vomiting	Nausea	Neither
Thiopentone	11	6	83
Methohexitone	14	12	74
Propanidid	25	18	57
Diazepam	5	3	92
Althesin	5	7	88
Ketamine	18	23	59
Etomidate	27	12	61
Diisopropylphenol	0	5	95

plement for nitrous oxide oxygen anaesthesia (see Table 6). When used in this way the i.v. anaesthetics fell into two groups. Those with a relatively low frequency of nausea and vomiting (e.g. thiopentone, 17%) tended to be agents with a moderately slow and smooth recovery; whereas, those with a high frequency of nausea and vomiting (e.g. ketamine, 41%) tended to have either rapid recovery characteristics or a high incidence of excitatory effects in the recovery period.

The incidence of nausea with intermittent thiopentone is less than that reported for a comparable technique with halothane (Dundee et al. 1965). Clarke (1984) points out that the high incidence of vomiting with propanidid and methohexitone is mostly "emergence vomiting" which, although less unpleasant than prolonged vomiting for the patient, may be more hazardous as it occurs immediately after wakening when protective reflexes may not have returned.

Although one would expect the use of intra-operative narcotic analgesics to be associated with a significant incidence of post-operative nausea and vomiting, little work has been done of a comparative nature in this area. In particular, it would be interesting to compare fentanyl with the new narcotic supplements alfentanil and sufentanil. In addition, the role of the relaxants and their reversal agents, given with atropine, a known anti-emetic, has been little studied. As these agents are necessary with high-dose opioid techniques, where a significant incidence of nausea and vomiting might be expected, they also serve to confuse the picture.

Pre-Medication

Pre-medication is a source of large variations in the reported incidence of post-operative nausea and vomiting. Unfortunately, the effect of pre-medication is frequently ignored. Many of the early studies are, thus, difficult to interpret, particularly when they report pooled data from large series.

Morrison et al. (1968) retrospectively analysed data from 6,700 patients undergoing either D and C or curettage only, and showed that they fell into three groups, depending on premedication (see Table 2) with constant differences for the two operations studied.

The contrasting influences of two components of pre-medication were shown in a carefully controlled prospective study by Riding in 1960. This remains one of the best-designed studies in this area, and the findings are unequivocal. The methodology used could be regarded as a blue-print for others examining the various aetiological and pharmacological influences on post-operative nausea and vomiting. Surprisingly, this lead has not so far been followed, with the result that a great deal of confusion still exists over many aspects of this area.

Riding examined the emetic effect of morphine and the possible anti-emetic effect of atropine as evidenced from data he obtained in an initial pilot study. He followed the pilot investigation with a carefully designed dose-response study, ensuring that each pre-medication group had an adequate number of patients (870 patients were divided into 12 groups). The trial involved double-blind random allocation of four doses of atropine (0, 0.3, 0.6, 1.2 mg) and three doses of morphine (0, 5, 10 mg) to women of child-bearing age having a standard operation (uterine curettage) under a standard anaesthetic (thiopentone, nitrous oxide, oxygen). Operations were carried out at the same time each day, and each patient was given the same level of post-operative care. The results demonstrate a clear dose-related emetic effect of morphine and a dose-related anti-emetic effect of atropine. Comparing the incidence of nausea, retching and vomiting, non-pre-medicated (saline) patients had an incidence of 22.4%, which was halved if 0.6 mg of atropine was used as pre-medication (incidence 11.5%). When, instead, 10 mg morphine was given as pre-medication, the incidence of nausea, retching and vomiting increased by a factor of 3 to 66.7%. If the morphine was combined with 0.6 m atropine, the incidence was again halved to 35.2%.

The other significant study of morphine/atropine pre-medication compared this regime with pethidine/atropine and with atropine alone (Dundee et al. 1965). Four different anaesthetic techniques were used involving nitrous oxide oxygen supplemented with halothane or cyclopropane (see Table 5) or intermittent thiopentone or intermittent propanidid in patients presenting for minor gynaecological operations. With atropine, most of the nausea and vomiting occurred in the first hour and less than 5% of patients were still vomiting after 1 h post-operatively. Addition of morphine to the pre-medication resulted in nausea and vomiting persisting for 6 h or more, and increased the percentage of vomiting with halothane from 10% to 43%. In contrast, pethidine, which is shorter acting than morphine, had its effects mainly in the first hour post-operatively and before the operation. Literature comparisons of pethidine and morphine as pre-medicants are contradictory (Palazzo and Strunin 1984a). Variations in the period of observation and dose-related effects may account for many of the differences. The dose-related effects of pethidine pre-medication were reported by Bellville (1961) and still require confirmation. At 1 mg/kg, pethidine inhibits nausea and vomiting. At lower doses there is no inhibition and at higher doses the emetic effect of pethidine predominates. This "biphasic" effect underlines the importance of dose in pre-medication and the need for dose-response studies to select the optimum dose for pre-medication.

Riding (1975) suggested that the high incidence of vomiting after minor operations with opioid pre-medication may be partly explained by patients becoming active early in the post-operative period and the opioid sensitising the patient to vestibular disturbance. This suggestion previously made by Purkis (1964) needs further study. Such early post-operative morbidity is clearly less likely after longer, more stressful operations, where recovery is somewhat prolonged.

The Patient and Surgery

The role of anaesthesia in post-operative nausea and vomiting is confused by the influence of the patient and the operation undergone. Not only are there direct effects for example in changes in the age of the patient, but also indirect effects as a result of the selection of particular anaesthetics to accommodate, for example, the elderly or particular operations (e.g. neurosurgery). Although the influence of patient weight has not been adequately studied, Bellville et al. (1960) suggested that obesity increased the incidence of sequelae, possibly due to longer desaturation times. In addition, a variety of non-anaesthetic factors over which the anaesthetist has little influence, have been reviewed by Palazzo and Strunin (1984a). These include metabolic, vestibular and psychogenic disturbances and gastro-intestinal disorders in particular are problematical, as well as intracranial disorders resulting in an increase in intracranial pressure in the semi-comatose patient.

Patient Predisposition

Although little studied, it is not unlikely that individuals show a wide variation in the vomiting threshold. Purkis (1964) suggested that sufferers from motion sickness suffer more than others from post-operative vomiting, particularly as opioids sensitise the vestibular centre. In his series of 1,713 patients, 1,015 had had a previous anaesthetic and of those who vomited previously, 42.7% vomited within 24 h on subsequent anaesthesia. In contrast, only 14.3% of those who did not vomit previously vomited on the second occasion. The remaining patients who had had no previous anaesthetic gave an incidence of vomiting of 30.2%.

Age

Children seem to vomit about twice as frequently as adults (Vance et al. 1973), although those less than 3 years old show a much lower incidence of vomiting (Rowley and Brown 1982). Palazzo and Strunin (1984a) suggest that some of the observed differences may be due to the use of different pre-medications. As adults increase in age, emetic sequelae become less of a problem (Burtles and Peckett 1957; Bellville et al. 1960), although Knapp and Beecher (1956) only found a significant decrease in nausea, retching and vomiting in patients aged more than 70 years. Howat (1960) and Purkis (1964) found no correlation of emetic sequelae with age, although Purkis tabulated several studies showing a clear age-related effect. He reviewed the complications of interpretation of these data by pointing out how the male : female ratio and the selection of anaesthetic change with age. Bodman et al. (1960) reported a higher percentage of fracture cases resulting in a higher incidence of vomiting in elderly patients, although old age by itself did not affect the vomiting rate.

As most of these studies were carried out in the period 1957–1960, Palazzo and Strunin (1984a) suggest that the time has come to repeat some of them with current techniques. Such an investigation would be useful as the number of elderly patients presenting for surgery has increased, thus, removing one of the problems with some of the earlier reports.

Sex

Women suffer two to three times more frequently than men from post-operative nausea and vomiting, and the various measures of severity indicate that the severity of vomiting is also similarly increased (Burtles and Peckett 1957); Bellville et al. 1960; Knapp and Beecher 1956). Purkis (1964) in his survey of 1,713 patients over 24 h post-operatively, rated both the incidence and severity of vomiting (see Table 7) and confirmed that women suffer more. Boulton (1955) pointed out that the difference between men and women persists independently of the type of anaesthetic and is seen when similar operations are performed on both sexes. Dyrberg (1962) showed that although women have a greater incidence than men of both nausea and vomiting following abdominal operations, when extra-abdominal operations are considered, the differences between the sexes are much smaller.

The differing incidence in severity between men and women is possibly due to hormonal factors as the highest incidence of nausea and vomiting in women occurs in the third and fourth week of menstrual cycle (Bellville 1961). In addition, there is no sex-related difference in children (Vance et al. 1973), at least until the age of 11 when differences start to appear (Smessaert and Schehr 1959).

The Operation

Palazzo and Strunin (1984a) underline the conflicting reports in this area due to a lack of adequately controlled data. None of the early studies are comparable, although the highest incidence of nausea and vomiting seems to be associated with intra-abdominal and ear nose and throat surgery. Each operation is per-

Table 7. Incidence and severity of nausea, retching and vomiting in men and women in the first 24 h following operation (Purkis 1964)

	Men ($n=437$)	Women ($n=1{,}276$)
Incidence	17.8%	33.2%
Severity score (number of episodes)	0.65	1.48

formed for a reason which in itself may distort patient selection. Indications often select age groups, and determine the ratio of male to female patients. It is clearly not justifiable to use pooled data from several operations to consider the influence of the operative site as different techniques and agents may be used for different age groups and age and anaesthetic may influence outcome.

As Purkis (1964) points out, if the same operation is carried out for different indications, then these may also include different age groups and may bias the comparison. He showed that intra-abdominal operations had more emetic sequelae than extra-abdominal operations, even when confined to one sex. Moreover, the influence of site was more important than duration of anaesthesia. Thus, appendicectomy operations which are relatively short have a higher incidence of nausea and vomiting than operations for varicose veins, which tend to last longer. Bonica and his colleagues (1958) noted that operations on the stomach, the duodenum and the gall bladder had a higher incidence (70%) than abdominal wall operations (15%). In a closely controlled study, Dundee et al. (1962) showed that D and C operations had a higher incidence of emetic sequelae than evacuation not requiring dilatation. Gynaecological operations frequently use emetic hormones, such as ergometrine, which in themselves could distort post-operative comparisons. Emergency operations, for example on accident victims, may also be a particular problem because of patients presenting with a full stomach.

The Post-Operative Period

Palazzo and Strunin (1984 b) pointed out that although there is a tendency in the post-operative period to rely on anti-emetics once vomiting has started, there are certain simple measures which are helpful in treating nausea and vomiting. These include treating post-operative pain and maintaining i.v. fluids to avoid hypotension. In addition, asking the patient to take deep breaths and to avoid movement is helpful. Finally, it is wise not to put patients close to those already vomiting.

Visceral or pelvic pain in the post-operative period is a common cause of nausea (Parkhouse 1963). Andersen and Krohg (1976) have shown that titration of patients with pain and nausea using opioids relieves both pain and nausea in 80% of episodes; whereas, opioids induced nausea in only 3.4% of cases. It is clear that careful post-operative use of opioids is a major beneficial factor in treating both nausea and pain successfully. It may be that some reports of nausea post-operatively in patients treated with opioids are as a result of the nausea returning with the pain rather than as a consequence of a direct opioid effect.

As motion sickness drugs are frequently effective in the treatment of post-operative nausea and vomiting, it seems likely that vestibular disturbances are an important influence in the post-operative period. Post-operative movement may be an important influence on opioid-induced emesis, as opioids sensitise the vestibular apparatus (Rubin and Winston 1958), and there is an increase in nausea and vomiting in ambulatory patients after morphine treatment when the patients are allowed to move around (Comroe and Dripps 1948). Bellville et al. (1960) suggested that periods of long post-operative sleep may be associated with a lower incidence of nausea and vomiting. Burtles and Peckett (1957) remarked that vomiting frequently follows the first oral intake of fluid or food. It is clear that much post-operative vomiting is a result of patients moving too early.

Clinical Research on Anti-Emetics

Mode of Action

Although the chemoreceptor trigger zone (CTZ) has a high density of dopaminergic receptors and is the main site of stimulus for drugs in the blood stream, anti-emetic agents with other than dopamine-blocking activity have been shown to be active in the treatment of post-operative nausea and vomiting. Most anti-emetics have some central activity on either the vomiting centre directly or the CTZ, and some accelerate gastric emptying although the significance of this is debatable. Although cyclizine (Brock-Utne et al. 1978) and domperidone and metoclopramide (Brogden et al. 1982; Pinder et al. 1976) have been shown to increase lower oesophageal sphincter pressure (LOSP) and, hence, may be useful in reducing reflux, the effective anti-emetic droperidol has been shown not to increase LOSP and this theoretically may increase the risk of reflux (Brock-Utne et al. 1978).

Domperidone, unlike metoclopramide, has difficulty crossing the blood-brain barrier and therefore its anti-dopamine effects are significant at peripheral sites only. The drug was designed (Reyntjens 1981) to influence the CTZ, which is outside the blood-brain barrier, without affecting the basal ganglia. To the extent that it has largely avoided CNS side-effects in patients with intact blood-brain barriers, this strategy has been successful.

Unfortunately, all of the agents described below have only limited efficacy by themselves, particularly in certain patients, and it may be that agents with different modes of action should be combined to block more than one pathway to achieve 100% success.

Methodology

Various studies have shown specific anti-emetic compounds to be superior to placebo, but as Bellville (1961) has pointed out, they do not help to identify the optimum agent. Comparisons are difficult, as studies are reported on different populations and usually no standard drug is included as a reference point. In addition, some investigators use methods so insensitive as not to be able to differentiate active compounds from placebo. Palazzo and Strunin (1984b) discussed the difficulties of selecting the most useful anti-emetic on the basis of current research data. Not only may effective prophylactics be inactive as treatments for active vomiting, but as the causes of vomiting are multi-factorial, no single agent may be ideal. Combinations of anti-emetics, although not yet investigated may reduce the incidence of unsuccessful treatment more effectively than higher doses of one agent. The use of more than one drug to treat post-operative nausea and vomiting in this area is attractive because of the possibilities of synergistic blockade of the different emetic impulses of the anaesthetic and surgical intervention.

As Clarke (1984) points out, it is not always possible to predict efficacy from the pharmacological nature of the compound being studied, and only well-controlled studies will provide reliable information. In this area, there is a clear need for standardisation. As most reports in the literature involve pooled operation data and observations from groups in which particular anaesthetics are given for particular operations, it is not surprising that there is little agreement between investigators. Most of the research on post-operative nausea and vomiting has been concentrated on the evaluation of different anti-emetics to the neglect of more basic aspects such as the frequency and severity of nausea and vomiting in different circumstances. The relevant influences have only been investigated as a secondary feature of drug evaluation.

The basic principle should be adopted of comparing the incidence and sometimes severity of nausea and vomiting in two or more comparable series of patients treated with different anaesthetics or drugs. Unfortunately, so great are the differences in detail in the methods employed in the literature, it is hardly ever possible to compare results from various reports. Thus, different studies have different definitions of vomiting, including or ignoring retching, and they may take no account of nausea. No allowance is made in some reports for amnesia of the patient for vomiting, an important post-operative feature. Unselected series provide large numbers of patients, but sexes and surgical procedures are often not equally represented and although these data may be useful to define the overall value of an anti-emetic, the obtaining of precise information from such series is much more difficult. There is a need for very careful control of patient selection, with stratified randomisation if necessary, as there is no other way of obtaining a clear understanding of the factors involved in post-operative vomiting and the efficacy of prophylactic and treatment regimes.

Bellville pointed out in 1961 that it is wise to use a statistician in the design of clinical trials on post-operative nausea and vomiting to ensure that there is a good chance of obtaining a meaningful result. It may be useful to consider pairing or matched data and scoring severity of subjective as well as objective responses. The scoring system and the terms employed need careful validation, and a pilot study should be performed to validate not only the methodology but also to select the appropriate doses of agents to be examined. The choice of dose of an anti-emetic is important. Too many studies use doses which are so high as to be on the plateau of the dose-response curve; whereas, a proper evaluation of the slope of the dose-response curve would allow a more accurate definition of the maximum useful dose and, hence, aid in the choice of the most effective antiemetic. Statistical advice is also required to define what numbers of patients would be adequate to compare two active treatments. From the number of "stand-offs" in the literature it is clear that this is done all too rarely. It should be remembered that although randomised double-blind trials are essential, these in themselves do not ensure that the populations in the treatment group are comparable and a check must be carried out at the end of the study to ensure that the populations are not biased as a result of a random variation in population selection. Where bias exists, it may be possible to allow for this in the analysis. Clear definition of the question to be asked and the population group to be examined will reduce these sorts of problems and improve the reliability of the study.

More though should be given to defining the question before the study is started. An illustration of a well-designed study from this point of view is the recent work from Abramowitz and co-workers (1983) which compared the prophylactic effect of droperidol and placebo on vomiting induced by strabismus surgery in children. A pilot study showed that 80% of such children vomited post-operatively and it was felt that a decrease of this figure to 50% would be clinically significant. A sequential trial was designed to see whether it was possible to obtain a statistically significant difference between placebo and droperidol at the 95% level, with a need to be 95% certain of detecting the change from 80% to 50%. The statisticians advised that this would require a maximum of 40 untied pairs of patients. Patients were then treated in a double-blind randomised fashion 30 min before the end of the operation with either 75 μg/kg droperidol or placebo. After 11 untied treatment pairs, it was found that 10 of these favoured droperidol compared with one favouring placebo and the statistically significant difference was demonstrated in the most efficient way.

The Anti-Emetics

Butyrophenones

Haloperidol and droperidol are powerful anti-dopamine agents with neuroleptic properties. Although repeat doses may cause extrapyramidal symptoms, hypotension and post-operative sedation, these effects are less severe than those observed with the phenothiazines (Palazzo and Strunin 1984b). Studies in volunteers have shown that these agents may result in motor restlessness and anxiety when they are given alone (Clarke 1984), although these problems seem rarely to be reported in patients following an operation.

Haloperidol is a potent prophylactic for vomiting in volunteers treated with apomorphine (Shields et al. 1971). Its prophylactic effect has been demonstrated in minor gynaecological operations without producing any significant effect on the recovery times from anaesthesia (Tornetta 1972). As a treatment for active vomiting induced by cyclopropane anaesthesia, haloperidol is clearly superior to placebo (Barton et al. 1975). The effectiveness of the agent for the treatment of nausea and vomiting was further demonstrated by Loeser and colleagues (1979) in a comparison of haloperidol and droperidol. They found that intramuscular haloperidol was effective as a treatment with rapid onset (less than 30 min), but with a relatively short duration of effect. Palazzo and Strunin (1984b) remark that it is difficult to understand why haloperidol is rarely used today as the agent has similar activity to the popular phenothiazines prochlorperazine and perphenazine, but it has fewer side effects. Perhaps the role of this agent should be reevaluated.

Loeser et al. (1979), comparing haloperidol and droperidol, reached the surprising conclusion that droperidol, in spite of its shorter half-life than haloperidol, had a longer action than haloperidol (up to 24 h). This, together with its undoubted efficacy, led Palazzo and Strunin (1984b) to settle on droperidol as their drug of choice for prophylaxis. Droperidol 5 mg intramuscularly is somewhat slower in onset than haloperidol, 2 mg (Loeser et al. 1979). The intravenous use of droperidol as a prophylactic is well documented (Patton et al. 1974; Iwamoto and Schwartz 1978; Kortilla et al. 1979; Rita et al. 1981; Abramowitz et al. 1983; Mortensen 1982; Santos and Datta 1984). In addition, the effectiveness of the drug as a treatment for vomiting has been demonstrated (Kortilla et al. 1979; Loeser et al. 1979). Kortilla and his colleagues (1979) noted that although large doses of droperidol have cardiovascular and sedative effects, the low doses (1.25 mg i.v.) used in their study lacked significant side effects.

Phenothiazines

Clarke (1984) points out that these drugs are particularly effective against agents acting directly on the CTZ and that higher doses may also depress the vomiting centre. The prototype, chlorpromazine, is now largely ignored in this area because of frequent undue sedation and hypotension.

Promethazine, another very sedative phenothiazine, is active as a prophylactic (Howat 1960) and in the treatment of active vomiting (Adriani et al. 1961). It should, however, only be used when sedation is required, as it is potentially hazardous for patients to vomit whilst sedated (Clarke 1984; Palazzo and Strunin 1984b). Promethazine is the most potent of the phenothiazines against motion sickness, and as it has a longer duration of action than hyoscine, it is the prophylactic of choice when motion sickness aetiology is involved, as for example in ear surgery (Palazzo and Strunin 1984b).

Prochlorperazine and perphenazine are the most popular agents of the phenothiazine group for the prevention and treatment of post-operative nausea and vomiting. They are equally effective against vomiting, although frequently are associated with extra-pyramidal symptoms (Palazzo and Strunin 1984b). As Clarke (1984) has pointed out, the extra-pyramidal symptoms liability of these drugs is cumulative and outlasts the anti-emetic action. This would ensure that if repeat doses are given, extra-pyramidal symptoms are more likely and when they occur can last for considerable periods.

Perphenazine is an effective prophylactic against morphine or pethidine induced vomiting, but is shorter acting than 10 mg morphine and has a significant incidence of extra-pyramidal symptoms. Lowering the dose reduces the restlessness found with this agent, but also reduces its effectiveness as an anti-emetic (Dundee et al. 1975). Loeser et al. (1979) showed that prochlorperazine was an effective treatment for post-operative vomiting, 10 mg intramuscularly acting between 30–60 min and having an effect for up to 4 h.

Anticholinergics

Anticholinergics, the oldest group fo anti-emetics, have been widely used in anaesthesia as anti-sialogogues and as their use declines with the newer anaesthetic techniques, it should not be forgotten that they have useful anti-emetic effects.

Clarke et al. (1965) compared hyoscine and atropine as anti-emetic pre-medication with 10 mg morphine. They concluded that hyoscine was superior to atropine because of its anti-emetic properties and its desirable sedative effect. The side effects of anticholinergics (dry mouth and sedation, and occasional disorientation in the elderly) have resulted in progressive restriction of hyoscine and atropine as pre-medicants.

Antihistamines

Palazzo and Strunin (1984b) suggest that because of its relative lack of side effects, cyclizine is the drug of choice for the treatment of active vomiting in spite of its short duration of action (4 h). Oculogyric crises have not been reported with cyclizine, but repeat injections can cause drowsiness and restlessness, although the incidence of side effects is less than the phenothiazines. Dundee and his colleagues (1975) preferred cyclizine to perphenazine as a prophylactic, although the duration of its effect was shorter than the emetic effect of morphine.

Metoclopramide

Although a relatively non-toxic agent, the prolonged use of metoclopramide has been associated with the occurrence of oculogyric crises, and normal doses given intravenously have caused other side effects (Clarke 1984; Hughes 1984). Palazzo and Strunin (1984b) point out that the efficacy of metoclopramide in either the prevention or treatment of post-operative nausea and vomiting remains to be proved. The agent is only effective prophylactically when given at the end of the operation. It has only poor anti-emetic effects when given with the pre-medication. Various studies on this drug have failed to identify a consistent long-lasting anti-emetic effect despite the use of varying doses and treatment regimes (Shah and Wilson 1972; Ellis and Spence 1970; Dundee and Clarke 1973; Assaf et al. 1974; Tornetta 1969). Even the demonstration of prophylactic efficacy, when given at the end of an operation (Handley 1967; Clark and Storrs 1969; Lind and Breivik 1970), has been contradicted (Dobkin et al. 1968), although this latter paper suffers from marked variations in the ratio of male to female patients.

Treatment has not been extensively studied, although Kortilla and colleagues (1979) compared prophylactic and therapeutic efficacies of metoclopramide, domperidone and droperidol, and concluded that metoclopramide and domperidone were no better than placebo.

Domperidone

Successfully designed as a peripherally acting dopamine blocker with only limited ability to cross the blood-brain barrier and hence avoiding CNS side effects, domperidone has been evaluated as a post-operative anti-emetic (Reyntjens 1981). The drug is felt by Palazzo and Strunin (1984b) to be more useful for treatment than for prophylaxis of nausea and vomiting. Data on the use of the drug as a prophylactic are contradictory (Boulanger et al. 1979; Wilson and Dundee 1979; Cooke et al. 1979). It may be that domperidone's relatively short duration of action or the selection of an inappropriately low dose or route of administration could explain this lack of activity. When used in the treatment of active vomiting, intravenous domperidone is clearly superior to placebo (Zegveld et al. 1978; Fragen and Caldwell 1978; Boghaert et al. 1980). Other studies of the drug given intramuscularly have failed to demonstrate an effect (Kortilla et al. 1979; Fragen and Caldwell 1979). Further work is needed on this agent with different dosage regimes to establish its suitability in prophylaxis as well as treatment.

Conclusions

Palazzo and Strunin (1984b) have reviewed the general measures which aid the patients likely to suffer from post-operative nausea and vomiting. Opioids are too useful to avoid, especially during the operation, and if used for pre-medication, may well benefit from the addition of a long-acting anti-emetic. A full stomach is best avoided although fasting is an unpredictable measure (Miller et al. 1983; Bodman et al. 1960). In situations such as emergencies, where danger from vomiting is acute, a rapid sequence induction and intubation remains the method of choice to avoid morbidity in patients with a full stomach. Other measures include avoiding gastric inflation, preventing blood reaching the stomach, for example after rhinoplasty, and avoiding excessive movement of the patient especially after opioids. Emergence from anaesthesia can produce gagging and possible vomiting by pharyngeal stimulation as protective reflexes return. Experienced anaesthetists would minimise this stimulus by using suction before reversal of relaxants and by removing the oral airway at the earliest opportunity. In addition, it is important to maintain pain relief, blood pressure and the hydration of the patient.

Adriani and colleagues in 1961 suggested that routine prophylaxis with anti-emetics was unjustified because of the low incidence of serious problems of vomiting and the high incidence of side effects with the then available anti-emetics. Palazzo and Strunin (1984b) modify this view and suggest that prophylaxis is justified in certain situations, for example oral surgery, where jaws are occluded by wires with a high risk of aspiration of vomitus, or in operations on the ear, eye or plastic operations to avoid the problems of disruption of delicate surgical work. In addition to these areas, there are patients with a high likelihood of vomiting who could also be protected. Sometimes it is difficult to suggest which patients should be so handled. Should, for example, all women be given anti-emetics? As Palazzo and Strunin (1984b) point out, if the operating room is a distant rough ride away and a woman has been given an opioid for an upper abdominal operation, then prophylaxis is not unreasonable.

Although Adriani et al. (1961) felt that most vomiting at emergence could be ignored, Clarke (1984) has pointed out that this could be more dangerous for the patient than prolonged vomiting, although not subjectively unpleasant. It would seem sensible to investigate methods of using the new anti-emetics with a low incidence of side effects (e.g. domperidone) by modifying dose recommendations where necessary. Thus, controlled release formulations or different dosage regimens may allow increased use of such agents in pre-medication or as prophylaxis. This would overcome Adriani's main objection to routine prophylaxis by the indiscriminate use of relatively toxic drugs. Indiscriminate use would still not be justifiable although the anaesthetist should certainly consider routine anti-emetic prophylaxis for susceptible groups, including women presenting for gynaecological operations, children, patients with a history of motion sickness or post-operative sickness, as well as patients presenting for oral, ear, eye, abdominal and plastic surgery.

The availability of new agents at least offers the anaesthetist more choices from which to judge how best to protect his patient.

References

Abramowitz MD, Oh TH, Epstein BS, Ruttimann UE, Friendly DS (1983) The antiemetic effect of droperidol following outpatient strabismus surgery in children. Anesthesiology 59:579

Adriani J, Summers FW, Antony SO (1961) Is the prophylactic use of antiemetics in surgical patients justified? JAMA 175:666

Andersen R, Krohg K (1976) Pain as a major cause of postoperative nausea. Can Anaesth Soc J 23:366

Assaf RAE, Clarke RSJ, Dundee JW, Samuel IO (1974) Studies of drugs given before anaesthesia. XXIV. Metoclopramide with morphine and pethidine. Br J Anaesth 56:514

Barton MD, Libonati M, Cohen PJ (1975) The use of haloperidol for treatment of postoperative nausea and vomiting – a double-blind placebo-controlled trial. Anesthesiology 42:508

Bellville JW (1961) Postanesthetic nausea and vomiting. Anesthesiology 22:773

Bellville JW, Bross IDJ, Howland WS (1960) Postoperative nausea and vomiting. IV. Factors related to postoperative nausea and vomiting. Anesthesiology 21:186

Bodman RI, Morton HJV, Thomas ET (1960) Vomiting by out-patients after nitrous oxide anaesthesia. Br Med J 1:1327

Boghaert A, Carron D, Gallant J, Stockman A (1980) Postoperative vomiting treated with domperidone. A double-blind comparison with metoclopramide and a placebo. Acta Anaesthesiol Belg 31:129

Bonica JJ, Crepps W, Monk B, Bennett B (1958) Postanesthetic nausea, retching and vomiting. Evaluation of cyclizine (marezine) suppositories for treatment. Anesthesiology 19:532

Borison HL, Wang SC (1953) Physiology and pharmacology of vomiting. Pharmacol Rev 5:193

Boulanger M, Dubois A, Lecron L (1979) Domperidone in the prevention of post-operative nausea and vomiting. Postgrad Med J 55 [Suppl 1]:45

Boulton TB, Chir B (1955) Oral chlorpromazine hydrochloride. A clinical trial in thoracic surgery. Anaesthesia 10:233

Brock-Utne JG, Rubin J, Welman S, Dimopoulos GE, Moshal MG, Downing JW (1978) The action of commonly used antiemetics on the lower oesophageal sphincter. Br J Anaesth 50:295

Brogden RN, Carmine AA, Heel RC, Speight TM, Avery GS (1982) Domperidone. A review of its pharmacological activity, pharmacokinetics and therapeutic efficacy in the symptomatic treatment of chronic dyspepsia and as an antiemetic. Drugs 24:360

Burtles R (1960) Post-operative vomiting: discussion. Proc R Soc Med 53:676

Burtles R, Peckett BW (1957) Postoperative vomiting. Br J Anaesth 29:114

Clark MM, Storrs JA (1969) The prevention of postoperative vomiting after abortion: metoclopramide. Br J Anaesth 41:890

Clarke RSJ (1984) Nausea and vomiting. Br J Anaesth 56:19

Clarke RSJ, Dundee JW, Love WJ (1965) Studies of drugs given before anaesthesia. VIII. Morphine 10 mg alone and with atropine or hyoscine. Br J Anaesth 37:772

Comroe JH, Dripps RD (1948) Reactions to morphine in ambulatory and bed patients. Surg Gynecol Obstet 87:221

Cooke RD, Comyn DJ, Ball RW (1979) Prevention of postoperative nausea and vomiting by domperidone. S Afr Med J 56:827
Coppolino CA, Wallace G (1962) Trimethobenzamide antiemetic in immediate postoperative period. JAMA 180:326
Costello DJ, Borison HL (1977) Naloxone antagonizes narcotic self-blockade of emesis in the cat. Pharmacol Exp Ther 203:222
Crocker JS, Vandam LD (1959) Concerning nausea and vomiting during spinal anesthesia. Anesthesiology 20:587
Cronin M, Redfern PA, Utting JE (1973) Psychometry and postoperative complaints in surgical patients. Br J Anaesth 45:879
Datta S, Alper MH, Ostheimer GW, Weiss JB (1982) Method of ephedrine administration and nausea and hypotension during spinal anesthesia for cesarean section. Anesthesiology 56:68
Dent SJ, Ramachandra V, Stephen CR (1955) Postoperative vomiting: incidence, analysis, and therapeutic measures in 3000 patients. Anesthesiology 16:564
Dobkin AB, Evers W, Israel JS (1968) Double-blind evaluation of metoclopramide, trimethobenzamide and a placebo as postanaesthetic anti-emetics following methoxyflurane anaesthesia. Can Anaesth Soc J 15:80
Dundee JW, Clarke RSJ (1973) The premedicant and anti-emetic action of metoclopramide. Postgrad Med J (July) [Suppl]:34
Dundee JW, Nicholl RM, Moore J (1962) Studies of drugs given before anaesthesia. III. A method for the studying of their effects on postoperative vomiting and nausea. Br J Anaesth 34:527
Dundee JW, Kirwan MJ, Clarke RSJ (1965) Anaesthesia and premedication as factors in postoperative vomiting. Acta Anaesthesiol Scand 9:223
Dundee JW, Assaf RAE, Loan WB, Morrison JD (1975) A comparison of the efficacy of cyclizine and perphenazine in reducing the emetic effects of morphine and pethidine. Br J Clin Pharmacol 2:81
Dyrberg V (1962) Haloperidol in the prevention of postoperative nausea and vomiting. Acta Anaesthesiol Scand 6:37
Ellis FR, Spence AA (1970) Clinical trials of metoclopramide as an antiemetic in anaesthesia. Anaesthesia 25:368
Fragen RJ, Caldwell N (1978) A new benzimidazole antiemetic, domperidone, for the treatment of postoperative nausea and vomiting. Anesthesiology 49:289
Fragen RJ, Caldwell N (1979) Antiemetic effectiveness of intramuscularly administered domperidone. Anesthesiology 51:460
Handley AJ (1967) Metoclopramide in the prevention of post-operative nausea and vomiting. Br J Clin Pract 21:460
Haumann JR, Foster PA (1963) The anti-emetic effect of halothane. Br J Anaesth 35:114
Heal PC (1965) Post-operative vomiting: Methoxyflurane and halothane. Anaesthesia 20:275
Howat DDC (1960) Anti-emetic drugs in anaesthesia. Anaesthesia 15:289
Hughes RL (1984) Hypotension and dysrhythmia following intravenous metoclopramide. Anaesthesia 39:720
Iwamoto K, Schwartz H (1978) Antiemetic effect of droperidol after ophthalmic surgery. Arch Ophthalmol 96:1378
Janhunen L, Tammisto T (1972) Postoperative vomiting after different modes of general anaesthesia. Ann Chir Gynaecol Fenniae 61:152
Knapp MR, Beecher HK (1956) Postanaesthetic nausea, vomiting and retching. Evaluation of the antiemetic drugs dimenhydrinate, chlorpromazine and pentobarbital sodium. JAMA 160:376
Korttila K, Kauste A, Auvinen J (1979) Comparison of domperidone, droperidol and metoclopramide in the prevention and treatment of nausea and vomiting after balanced general anesthesia. Anesth Analg 58:396
Kvisselgaard N (1958) Chlorpromazine and chlorcyclizine in the prevention of postoperative nausea and vomiting. Acta Anaesthesiol Scand 2:153
Lind B, Breivik H (1970) Metoclopramide and perphenazine in the prevention of postoperative nausea and vomiting. Br J Anaesth 42:614

Loeser EA, Bennett G, Stanley TH, Machin R (1979) Comparison of droperidol, haloperidol and prochlorperazine as postoperative anti-emetics. Can Anaesth Soc J 26:125
Miller M, Wishart HY, Nimmo WS (1983) Gastric contents as induction of anaesthesia. Is a 4 hour fast necessary? Br J Anaesth 55:1185
Morrison JD, Hill GB, Dundee JW (1968) Studies of drugs given before anaesthesia. XV. Evaluation of the method of study after 10000 observations. Br J Anaesth 40:890
Mortensen PT (1982) Droperidol postoperative anti-emetic effect when given intravenously to gynaecological patients. Acta Anaesthesiol Scand 26:48
Nimmo WS (1984) Effect of anaesthesia on gastric motility and emptying. Br J Anaesth 56:29
Palazzo MGA, Strunin L (1984a) Anaesthesia and emesis. I. Etiology. Can Anaesth Soc J 31:178
Palazzo MGA, Strunin L (1984b) Anaesthesia and emesis. II. Prevention and management. Can Anaesth Soc J 31:407
Parkhouse J (1963) The cure for postoperative vomiting. Br J Anaesth 35:189
Parkhouse J, Henrie JR, Duncan GM, Rome HP (1960) Nitrous oxide analgesia in relation to mental performance. J Pharmacol Exp Ther 128:44
Patton CM, Moon MR, Dannemiller FJ (1974) The prophylactic antiemetic effect of droperidol. Anaesth Analg 53:361
Pinder RM, Brogden RN, Sawyer PR, Speight TM, Avery GS (1976) Metoclopramide: a review of its pharmacological properties and clinical use. Drugs 12:81
Purkis IE (1964) Factors that influence postoperative vomiting. Can Anaesth Soc J 11:335
Ratra CK, Badola RP, Bhargava KP (1972) A study of factors concerned in emesis during spinal anaesthesia. Br J Anaesth 44:1208
Reyntjens AJ (1981) The anti-emetic action of domperidone – a review. R Soc Med Int Congr Symp Ser No 36:97
Riding JE (1960) Post-operative vomiting. Proc R Soc Med 53:671
Riding JE (1963) The prevention of postoperative vomiting. Br J Anaesth 35:180
Riding JE (1975) Minor complications of general anaesthesia. Br J Anaesth 47:91
Rita L, Goodarzi M, Seleny F (1981) Effect of low dose droperidol on postoperative vomiting in children. Can Anaesth Soc J 28:259
Rowley MP, Brown TCK (1982) Postoperative vomiting in children. Anaesth Intens Care 10:309
Rubin A, Winston J (1958) The role of the vestibular apparatus in the production of nausea and vomiting following the administration of morphine to man. J Clin Invest 29:1261
Santos A, Datta S (1984) Prophylactic use of droperidol for control of nausea and vomiting during spinal anesthesia for cesarean section. Anesth Analg 63:85
Shah ZP, Wilson J (1972) An evaluation of metoclopramide (Maxolon) as an anti-emetic in minor gynaecological surgery. Br J Anaesth 44:865
Shields KG, Ballinger CM, Hathaway BN (1971) Antiemetic effectiveness of haloperidol in human volunteers challenged with apomorphine. Anesth Analg 50:1017
Smessaert A, Schehr CA (1959) Nausea and vomiting in the immediate postanaesthetic period. JAMA 170:118/2072
Smith BL, Manford MLM (1974) Postoperative vomiting after paediatric adenotonsillectomy. Br J Anaesth 46:373
Stein JM (1982) Factors affecting nausea and vomiting in the plastic surgery patient. Plast Reconstr Surg 70:505
Tornetta FJ (1969) Clinical studies with the new antiemetic metoclopramide. Anesth Analg 48:198
Tornetta FJ (1972) Double-blind evaluation of haloperidol for antiemetic activity. Anesth Analg 51:964
Vance JP, Neill RS, Norris W (1973) The incidence and aetiology of post-operative nausea and vomiting in a plastic surgical unit. Br J Plast Surg 26:336
Waters RM (1936) Present status of cyclopropane. Br Med J 2:1013
Wilson DB, Dundee JW (1979) Evaluation of the anti-emetic action of domperidone. Anaesthesia 34:765
Zegveld C, Knape H, Smits J, Belopavlovic M, Caron D, Gallant J, Stockman A, Boghaert A (1978) Domperidone in the treatment of postoperative vomiting: a double-blind multicenter study. Anesth Analg 57:700

Mechanisms and Treatment of Nausea and Vomiting in Pregnancy

D. V. I. Fairweather[1]

Introduction

Nausea or vomiting in the early morning, so-called "morning sickness", is so common that it is accepted as a symptom of normal pregnancy. Occurring soon after waking, it is often retching rather than actual vomiting and usually does not disturb the woman's health or her pregnancy and the symptoms nearly always stop before the 14th week. In a much smaller proportion of cases, probably now not more than 1 in 1,000 total births, the vomiting becomes more serious and persistent, occurring throughout the day and even during the night. To this serious form of vomiting, the term hyperemesis gravidarum is given. It is difficult to find a definition of hyperemesis gravidarum in any standard obstetrics textbook and naturally one is faced with the problem: "when does the vomiting in pregnancy become serious enough to warrant the label hyperemesis?". In 1956, the American Council on Pharmacy and Chemistry gave a lengthy definition restricting the term "nausea and vomiting of pregnancy" to the first 14–16 weeks and to cases showing no signs of disturbed nutrition. The term hyperemesis gravidarum was reserved for those few patients developing intractable vomiting and disturbed nutrition, with upset of electrolyte balance, weight loss of more than 5% or other more severe signs such as neurological disturbances, and liver or renal damage.

For the purpose of this discussion and for my personal studies, I have defined hyperemesis gravidarum as "Vomiting occurring in pregnancy, appearing for the first time before the twentieth week of gestation and of such severity as to require the patient's admission to hospital, the vomiting being unassociated with such coincidental conditions as appendicitis, pyelitis, etc.". This definition too has its flaws for universal use, (e.g. the differences between areas regarding police and the availability of hospital beds), but it gives a good working basis and sets certain limitations for diagnosis. For Examples, are time of onset, and exclusion of other possible causes of vomiting, which are often not made clear. Vomiting in the latter moths of pregnancy seems to be unrelated to the early vomiting and, indeed, some authors doubt whether there is a separate disease entity due to pregnancy that causes excessive vomiting in the latter months. Usually, other underlying conditions are found, e.g. pyelonephritis and pre-eclampsia. More often, mechanical factors, e.g. multiple pregnancy or hydramnios causing distension with backflow of gastric contents though the lax cardiac sphincter of the stomach, and hiatus hernia, produce late vomiting associated with heartburn. In this discussion I do not intend to deal with this late vomiting.

1 Department of Obstetrics and Gynaecology, Faculty of Clinical Sciences, University College Hospital School of Medicine, University of London, 88–96 Chenies Mews, London WC1E 6HX, UK

Nausea and Vomiting: Mechanisms and Treatment
Ed. by C. J. Davis, G. V. Lake-Bakaar and D. G. Grahame-Smith

Aetiology

Although the title is given as including the mechanism of vomiting in pregnancy, I intend to discharge this responsibility by addressing the question of aetiology. As you are all probably aware, there has been a profusion of theories of aetiology over the years, indicating that we really do not have any one satisfactory or universally acceptable theory. Historical references in the literature to vomiting in early pregnancy go back to 2,000 BC and contain many and varied suggestions, some quite bizarre, for the alleviation of this condition. From the early writings it was clear that mild sickness and nausea was thought to be seldom of dangerous consequence and to be spontaneously relieved by about the end of the 4th month of gestation. According to Williams (1906), Kerkring in 1706 was the first to report the death of a pregnant woman from "icterus gravis" associated with vomiting. Williams notes that Simmons in 1813 first induced abortion for the relief of hyperemesis gravidarum. However, it was not until 1852 that the justifiability of this procedure was first evidently brought to the attention of the profession in a discussion before the Academy of Medicine of Paris, where Dubois and Danyau contended that induction of abortion was not only justifiable but urgently demanded in severe cases of vomiting, while Cazeaux held that interference was unnecessary and even hastened death in some instances. It was during that debate that Dubois (1852) gave the first classical clinical description of pernicious vomiting during pregnancy, which he divided into three stages:

> First, constant vomiting associated with considerable emaciation and frequently with troublesome dribbling of saliva, at the same time urine becomes scanty and highly coloured and the pulse somewhat accelerated. Second, the vomiting becomes still more severe, emaciation more pronounced and the breath assumes a peculiar acid and fetid odour. The pulse increases in frequency and there may be a slight rise in temperature. Third, there is an apparent amelioration of symptoms for a short time during which vomiting sometimes ceases and the patient passes into delirium or a torpid condition and dies in coma or convulsions.

By now it was clear that many authors were becoming interested in the aetiology of hyperemesis gravidarum and the early theories were considered in three main groups:

1. Reflex, related to things such as displacements of the uterus, endometritis, ovarian tumours or abnormalities of the ovum, for example hydramnios, hydatidiform mole and twin pregnancy
2. Neurotic, as suggested by Kaltenbach in 1891 to the Berlin Obstetrical Society, when he concluded that the vomiting of pregnancy is usually a manifestation of neurosis, somewhat allied to hysteria and readily amenable to suggestive treatment
3. Toxaemic, where it was postulated that there was some circulating toxin arising either from the gastro-intestinal tract, the ovum and appendages, ovarian secretions or hepatic lesions.

More recently, however, we have tended to group theories under four main headings: endocrine, psychosomatic, allergic and metabolic.

Endocrine Factors

Earlier authors postulated that the problem was related to the secretion of the corpus luteum or to some other ovarian factors and suggested that it was significant that symptoms usually disappeared at about the time when retrogression of the corpus luteum began. This led to numerous investigations of oestrogen and progesterone levels in patients suffering from excessive vomiting. Some used oestrogens to treat hyperemesis, arguing that they would depress anterior pituitary function, decreasing luteinising hormone and, thus, decreasing the secretion of the corpus luteum. Others, however, were unable to demonstrate any relation between oestrogen levels and the degree of nausea and vomiting. The Smiths and Hurwitz, however, thought that oestrogens per se were not responsible for the progesterone-stimulating effects of oestrogen, but that this was effected through increased utilisation of chorionic gonadotrophin (Smith et al. 1944). Gradually, therefore, the interest in oestrogen and progesterone as aetiologic factors waned and my own studies also confirmed lack of relation between circulating levels of those hormones and the occurrence of vomiting during pregnancy. However, a continuing interest in the role of chorionic gonadotrophin has been maintained. Brindeau in 1937 reported higher than normal titres of the gonadotrophic hormone, prolan B in hyperemesis and a few years later Way noted a high incidence of hyperemesis in patients suffering from hydatidiform mole, suggesting that high levels of anterior pituitary-like hormone associated with this condition were responsible for the low acid secretion in the gastric juice, which in turn was related to the vomiting (Way 1945). Schoeneck at about the same time presented evidence of increased levels of gonadotrophic hormone in the urine and serum of patients suffering nausea and vomiting as compared with those who were free from symptoms (Schoeneck 1942). Attention was also drawn to the time relation between the highest levels of chorionic gonadotrophin and the occurrence of pregnancy vomiting, the greatest concentration of HCG being present normally from 6 to 12 weeks of pregnancy, the time when hyperemesis was most common. I also found the peak incidence of symptoms to be between 6 and 12 weeks.

Fairweather and Loraine then went on in a clinical study to compare the urinary HCG excretion in hyperemesis patients with that in a series of normal pregnant women at comparable stages of pregnancy and found that in patients with hyperemesis the mean HCG excretion between the 7th and 15th weeks was significantly lower than that found in normally pregnant women – evidence, of course, conflicting with that of Schoeneck (Fairweather and Loraine 1962). Fairweather further investigated both blood and urine levels of HCG in hyperemesis including the calculation of clearance rates and found results comparable to findings in normal pregnancy (Fairweather 1965). From all the evidence, therefore, it seems unlikely that either high levels of HCG or undue sensitivity to normal levels can be blamed for the occurrence of hyperemesis. One other endocrine aspect has interested authors from time to time, namely, the function of the anterior pituitary-adrenal cortex axis in hyperemesis and some have been impressed with the value of ACTH in the treatment of hyperemesis, supporting the view that there was hypofunction of the adrenal cortex. More recent studies, however, showing raised level suggest hyper-rather than hypofunction.

Psychosomatic Factors

I have already quoted Kaltenbach's view that the aetiology is neurosis allied to hysteria, and indeed, this is thought to explain the similar symptoms in the husbands in some of those patients. Other authors have suggested different neurotic factors such as repudiation of femininity or a conscious or unconscious wish not to become pregnant, or even a form of self-punishment. Robertson describing a dyspeptic syndrome noted that a high proportion of these women in marriage were unduly attached to their mothers and that in due course they suffered from nausea and vomiting during pregnancy (Robertson 1946). The maternal attachment was also often associated with a disturbed sexual function in the patient. Bertling (1948) suggested a common factor, probably psychogenic, between dysmenorrhea and nausea and vomiting of pregnancy and other authors in a detailed pathologic study of patients suffering from hyperemesis noted that these patients were emotionally unstable and sexually frigid and had nearly all suffered from severe dysmenorrhea. Their personality features showed general psychological immaturity. Some studies have investigated the association between clinical psychiatric disease and hyperemesis and found that the only chronic psychiatric illness that appears significantly more frequently in women in hyperemesis was hysteria, 15% compared with 2% of controls. Not all authors, however, agreed, although there do seem to be some strong points in favour of a psychologic basis of the vomiting and one is tempted to think that the success claimed for the many varied methods of treatment may be more the result of psychotherapy than of the drugs prescribed. One study carried out by myself to investigate the part played by psychiatric factors in hyperemesis used a variety of standard psychiatric and personality tests on hyperemesis and control patients (Fairweather 1965). The findings confirmed that the majority of the hyperemesis patients had infantile or immature personalities associated with hysteria and, furthermore, these patients showed evidence of strong maternal dependance. There seems little doubt that psychiatric factors are implicated in a proportion of cases of hyperemesis and I am of the opinion that this proportion is as high as 70% to 80% of all cases.

Allergic Factors

The implication of allergy in hyperemesis was promoted by Hofbauer's suggestion in 1926 that histamine poisoning was a factor and Kappeler-Adler claimed that histamine could be isolated in considerable amounts in the urine of patients with hyperemesis (Kappeler-Adler 1941). The allergy theory gained further support from various workers who noted that in many allergic disorders nausea and vomiting became a feature at one stage or another. Finch (1940) thought that the allergic sensitivity was to the secretions of the patient's own corpus luteum and Johnson, in the same year (1940), suggested that the allergen might be a specific protein substance of which the syncytial cells could be the chief component. If they entered the maternal circulation sensitivity to foreign proteins could reproduced. He attempted to densensitise patients using husband's blood.

This was also attempted by other workers who believed that the symptoms of hyperemesis were caused by isoaglutinins as in erythroblastosis. From an historical point of view, however, Fieux in 1912 first suggested this aetiologic factor and therapy when he claimed that cure frequently followed the injection of 10 to 20 ml serum obtained from normal pregnant women. The sensitising agent was claimed by James (1945) to be the semen of the spouse, and this view was also expressed later by Dougray (1949). Randolph and Rollins (1950) suggested that the basic mechanism of allergic disease was related to the function of the pituitary adrenal system. Adreno-cortical insufficiency causing hypersensitivity to histamine, favouring the development of allergic manifestations and this was suggested again in 1955 by Youssef and Staemmler in relation to the importance of allergy and adreno-cortical function in the aetiology of hyperemesis. Although many claims have been made in the past for the success of treatment using antihistamines or their derivatives, no specific allergic factor has ever been satisfactorily demonstrated and one must again consider the possibility that the good results were due to psychological effect, a coincidental sedative action or both.

Metabolic Factors

In 1918 Duncan and Harding first advanced the theory of carbohydrate deficiency or glycogen deficiency of the liver as the cause of the nausea and vomiting of pregnancy and since then many others have referred to similar factors. Whether this deficiency was primary or resulted from the vomiting was questioned. It was always clear, however, that dehydration and ketosis, which were features of severe vomiting, were secondary. Vara and Vehniainen (1949) reviewing the aetiology and treatment of pregnancy vomiting stressed the importance of fluid balance, but it was Bergqvist who drew attention to the importance of the potassium ion pointing out that lack of potassium, primarily intracellular, is a significant factor in many diseases in which there is disturbance of fluid and ion balance and that in some cases it might be the cause of death (Bergqvist 1951). Lans et al. (1953) and others also recognised the danger of hypokalaemia and it is clear that many of the deaths from hyperemesis previously attributed to toxaemia were a result of potassium depletion causing respiratory or cardiac failure. Overall, however, it seems more likely that metabolic changes are secondary to vomiting during pregnancy rather than a primary aetiologic factor.

Incidence

Turning now to the incidence of the condition, it has been reported frequently that there are differences between countries and within various countries. The popular suggestion was that the condition was common in western civilizations and almost unknown amongst Eskimos and native African tribes. Changing incidences within a single centre or country have been noted and many studies have shown low incidences in times of war. The reason for this has been variously suggested as diet, different psychological or psycho-social environment or purely psychologic factors. There seems little doubt, however, that severe vomiting during pregnancy is now relatively uncommon with an incidence probably as low as 1 per 1,000 total births, as I have already mentioned. Some epidemiologic factors have been suggested as relevant to differences in incidence of the condition. Multiple pregnancy has often been suggested as a factor and high levels of HCG found in twin pregnancy were thought to be relevant. In my own study of over 90,000 single and almost 1,000 multiple pregnancies, I found a highly significant relationship between multiple pregnancy and hyperemesis (Fairweather 1965). Most authors have reported that parity has no influence on incidence of hyperemesis, although some have suggested an increased incidence in primigravidae. My study found that incidence was higher in first pregnancies when a white only population was considered, but in the non-white population there was no significant difference between the incidence in first and later pregnancies. Race and colour do not significantly influence the incidence of hyperemesis and contrary to expectation, there is no evidence that marital status has any affect on the occurrence of hyperemesis. Indeed, I found a negative correlation with respect to both illegitimate pregnancy and pre-nuptial conception. Hyperemesis tends to recur in subsequent pregnancies, figures from 26% to 50% being quoted for patients in their second or subsequent pregnancy who require admission to hospital because of hyperemesis and have a history of admission in the previous pregnancy with a similar diagnosis. Few authors have investigated the relationship between previous history of unsuccessful pregnancy and hyperemesis, although one author noted that the incidence of hyperemesis in multiparous patients who had previously aborted was 19% greater than in the remaining obstetric population. In my series of hyperemesis patients over 40% had a history of previous unsuccessful pregnancy, abortion, stillbirth or neonatal death, suggesting that this prior history might be an important aetiological factor, possibly operating through a psychological pathway.

Effects of Hyperemesis Gravidarum on the Outcome of Pregnancy

Many authors have studied the outcome of pregnancy and the effect of hyperemesis. Early writers suggested an increased incidence of spontaneous abortion compared with normal obstetric patients, but recent studies, including my own, have failed to confirm this and, indeed, some have suggested that patients without nausea and vomiting are more likely to have the problem. My study was in fact the first to demonstrate that hyperemesis has no effect on the birth weight of the infant and this has since been confirmed by other workers (Fairweather 1968). As you can imagine, the literature concerning hyperemesis as a cause of foetal deformity is difficult to evaluate in terms of quoted rates when the types of deformities are not detailed and the problem was further compounded in the 1950s and 1960s when various anti-emetic drugs were being widely used in treatment and the possible teratogenic effect of some of these was questioned. Prospective studies, including my own, have, however, confirmed that there is little or no evidence that hyperemesis per se causes an increased risk of deformity in the child. Hyperemesis used to be a potentially lethal condition, as far as the mother was concerned, as shown by the maternal mortality rates (159 per million births) quoted for England and Wales from the 1930s. By the 1950s, however, the pattern had changed (3 per million births) and it is clear that this was due to the increased understanding of fluid and electrolyte balance and in particular to the demonstration and active correction of extreme hypokalaemia. The understanding of the importance of fluid and electrolyte balance resulted in earlier and more efficient treatment.

Treatment

Finally, I would like to address the question of treatment. In medical conditions in which the cause is unknown it is usual to find a wide variety of suggested therapies, and nausea and vomiting during pregnancy and hyperemesis gravidarum are no exception. My survey of treatment methods (Fairweather 1978) listed 30 used in as many years up to 1968. In subsequent years, however, suggested therapy has been mainly drugs of the anti-emetic variety, but during this time because of the thalidomide tragedy there has been an understandable reluctance to use drugs of any kind during early pregnancy. Only in July of 1984 was this aspect further highlighted by the decision of a major US drug company to meet the claims of patients alleging that the drug Debendox (dicyclamine, doxylamine and pyridoxine) taken in early pregnancy has been responsible for causing foetal malformations. Probably the only value of drug therapy is at the stage of morning sickness when anti-emetics or mild sedatives may counter the feeling of nausea and prevent the women from developing excessive vomiting and entering the vicious circle of dehydration, starvation and electrolyte imbalance. Once the patient has reached the stage of hyperemesis gravidarum much more basic therapy is required, and the regimen calls for correction of dehydration, carbohydrate deficiency and ionic deficiencies. This is best managed by intravenous therapy with or without the addition of vitamin supplements and sedative agents. The need for and value of psychotherapy must be assessed in each case. The following outlines the principle of treatment which I have found to be helpful. It is best to admit patients to a small ward, and rarely do they benefit from single accommodation. They should be confined to bed in the initial phase and have routine observations performed. The history should be taken from both the patient and from relatives. Measurements of weight, urine testing (particularly bacteriology of the urine to exclude urinary infection), maintenance of fluid balance charts and recording of haemoglobin and electrolytes are essential. If the patient is dehydrated, intravenous therapy should be given to correct this and to obtain control of electrolyte balance. Twice daily blood chemistry is advisable, and the patient should have nothing by mouth at that stage. If, on the other hand, the patient is not dehydrated then oral, plain water in small amounts, hourly, is the best therapy. The patient should also be sedated. Further therapy should obviously depend on the response to treatment, and when oral fluids are tolerated, the amount can be increased and the type varied so that gradual return to full diet is obtained.

The patient should be mobilised as soon as possible. One important thing is to keep a check on weight gain; I have found that the best guide to recovery is found in the patient's weight chart. Static or falling weight before discharge should alert attendants to the possibility of rapid recurrence of the symptoms. In my own experience, adhering to the above principles, I have had satisfactory results in all cases and in no instance has termination of pregnancy been necessary.

King in 1955 suggested that the successful treatment of nausea and vomiting of pregnancy could be expressed by the formula:

$$\text{Cure} = \frac{T + U + X}{PH + PS},$$

where T represents the passage of time which is generally the largest single favourable factor in the formula; U is understanding of the patient by the physician and his approach to her, most often it is in the nature of sympathy, support and assurance, but it may be firm and disciplinary on occasion; this is always time-consuming for the doctor, but is his most effective weapon to date; X represents medication and may be one of the sedative or anti-emetic drugs or a placebo (ideally, X equals some substance yet undiscovered which directly and scientifically cancels out the PH factor in the denominator); PH is equal to primary physiological derangement in pregnancy, nature as yet unknown; and PS stands for adverse psychological factors.

The cure is successful if the sum of all elements in the numerator is larger than the sum of the elements in the denominator. I agree with King that many factors have to be taken into consideration and that at all times the patient must be handled firmly, but kindly, and should be allowed to discuss her troubles freely with the doctor and members of the nursing team. In some cases in which there is an obvious psychological problem the help of a psychiatrist may be required, but this is usually not necessary.

References

American Council on Pharmacy and Chemistry (1956) Current status of therapy in nausea and vomiting of pregnancy. JAMA 160:208–209

Bergqvist N (1951) Potassium deficiency in hyperemesis gravidarum. Acta Obstet Gynecol Scand 30:428–438

Bertling MH (1948) Some psychic aspects of dysmenorrhoea and nausea and vomiting of pregnancy. Am J Obstet Gynecol 56:733–737

Brindeau A, Hinglais H, Hinglais M (1937) Vomissements incoercibles de la grossesse et hormone gonadotrope. C R Soc Biol (Paris) 124:349–351

Dougray T (1949) Antihistamines in the treatment of nausea and vomiting in pregnancy. Br Med J 2:1081

Dubois P (1852) Bull Acad Med (Paris) 17:557

Duncan JW, Harding VJ (1918) A report on the effect of high carbohydrate feeding on the nausea and vomiting of pregnancy. Can Med Assoc J 7:1057

Fairweather DVI (1965) Hyperemesis gravidarum. MD Thesis, University of St. Andrew's, Scotland

Fairweather DVI (1968) Nausea and vomiting in pregnancy. Am J Obstet Gynecol 102:135–175

Fairweather DVI (1978) Nausea and vomiting during pregnancy. Obstet Gynecol Annu 7:91–105

Fairweather DVI, Loraine J (1962) Urinary excretion of human chorionic gondadotrophin in patients with hyperemesis gravidarum. Br Med J 1:666–669

Fieux J (1912) La serotherapie appliquée au traitement des vomissements graves de la gestation. Ann Gynecol Obstet 9:718

Finch JW (1940) The nausea and vomiting of pregnancy due to allergic reaction – a study of 192 cases. Am J Obstet Gynecol 40:1029–1036

Hofbauer J (1926) Experimental studies on the toxemias of pregnancy: can histamine poisoning be regarded as the etiologic factor? Am J Obstet Gynecol 12:159–189

James DW (1945) Pernicious vomiting of pregnancy due to sensitivity to semen. West J Surg 53:380–382

Johnson HW (1940) Toxaemias of pregnancy. Surg Gynecol Obstet 70:513–516

Kapeller-Adler R (1941) Histidine metabolism in toxaemia of pregnancy. Biochem J 35:213–218

Kaltenbach R (1891) Über Hyperemesis gravidarum. Z Geburtshilfe Gynaekol 21:200

King AG (1955) The treatment of pregnancy nausea with a pill. Obstet Gynecol 6:332–338

Lans HS, Gollin HA, Daro AF, Noral EJ (1953) Hypokalemia due to persistent vomiting during pregnancy. JAMA 153:1012–1015

Randolph TG, Rollins JP (1950) Relief of allergic diseases by ACTH therapy. In: Mote JR (ed) Proceedings of the First Clinical ACTH Conference, London, Churchill, London, pp 479–490

Robertson GG (1946) Nausea and vomiting of pregnancy: study in psychosomatic and social medicine. Lancet 2:336–341

Schoeneck FJ (1942) Gonadotropic hormone concentration in hyperemesis gravidarum. Am J Obstet Gynecol 43:308–312

Smith OW, Smith GVS, Hurwitz D (1944) The relationship between hormonal abnormalities and accidents of late pregnancy in diabetic women. Am J Med Sci 208:25–35

Vara P, Vehniainen K (1949) Fluid balance in hyperemesis gravidarum. Ann Chir Gynaecol Fenn [Suppl 3] 38:540–556

Way S (1945) Relation between gastric acidity and anterior pituitary-like hormone content of urine in pregnant women. Br Med J 2:182–184

Williams JW (1906) Pernicious vomiting of pregnancy. Johns Hopkins Hosp Bull 17:71–92

Youssef AJ, Staemmler HJ (1955) The relation between allergy and adrenocortical function and its significance in the aetiology of hyperemesis. Acta Endocrinol 18:109–115

A Neurologist's View of Nausea and Vomiting

J. D. PARKES [1]

Introduction: Vomiting is Common in Neurological Disorders

The mechanisms (and treatment) of sickness depend on the cause. Some of the neurological causes of vomiting are the following (see also Table 1):

1. Migraine. Classic migraine attacks are characterised by nausea, and also by gastric stasis. Intense pain may contribute to gastric stasis, which may be the cause of the sensation of nausea, and may or may not lead to vomiting. Moreover, it is possible that the effect of analgesics is increased by administration of an anti-emetic.
2. Raised intracranial pressure. Vomiting may be sudden and forceful, but more often is preceded by severe nausea. Relief of raised pressure (e.g. surgery, dexamethasone) relieves sickness within a few days.
3. Low brain-stem focal gliomas. These may cause severe and persistent vomiting, with or without nausea, in the early stages, and without raised intracranial pressure, pointing to a vomiting centre in man localized in the low medulla.
4. Vestibular disorders. Vomiting is a common feature in ear disease, Meniere's disease, vestibular neuronitis. Anticholinergic and antihistaminic, but not antidopaminergic, drugs give relief.

Table 1. Common causes of sickness in clinical practice

Example	Possible mechanism	Treatment employed
Digitalis Levo-dopa Apomorphine	Stimulation of chemo-receptor mechanisms	Dopamine antagonists
Meniere's disease	Stimulation of labyrinthine mechanisms	Anticholinergic drugs
Anxiety, high emotion	Cortical stimulation	Benzodiazepine
Cerebral haemorrhage, tumour	Raised intracranial pressure	Prednisolone
Pregnancy Gastro-enteritis	Stimulation of abdominal afferents	Pyridoxine
Migraine	Central and gastric effects	Gastrokinetic antagonist, e.g. domperidone, metoclopramide

1 Department of Neurology, King's College Hospital and Institute of Psychiatry, University of London, Denmark Hill, London SE5, UK

Nausea and Vomiting: Mechanisms and Treatment
Ed. by C. J. Davis, G. V. Lake-Bakaar and D. G. Grahame-Smith

Table 2. Approximate frequency of vomiting with different anti-parkinsonian drugs

Drug	Approximate frequency of vomiting in initial treatment (in conventional dosages) %
Oral levo-dopa	80
Oral levo-dopal with decarboxylase inhibitor	15
Oral amantadine	10
Oral bromocriptine	30
Oral pergolide	30
Intravenous lisuride	50
Oral transdihydrolisuride	15
Anticholinergic drugs (mainly oral)	5

Data from King's College Hospital Parkinson's Disease Clinic. In most instances, tolerance to emetic effect occurs after 1–6 months.

5. Dopamine-like drugs. All the dopamine agonists used to treat Parkinson's disease may cause vomiting, but the emetic properties do not always parallel the motor effects (see Table 2). Nausea and vomiting may be due, at least experimentally to gastric effects on stomach dopamine systems, not to postulated stimulation of a chemoreceptor zone in the brain stem. Could it be that the major chemoreceptor trigger area in man lies in the stomach, not the brain?

Some neurologists, by the time of reaching specialist status, have little contact with acute medicine, and may not have witnessed a patient vomit for over a decade. Faced with this dilemma, the wise physician employs Henry Miller's famous dictum applied to giddiness – "leave it to the house physician" (Matthews 1963).

Neurological Causes of Vomiting

Drug-Induced Sickness

I well remember my first encounter with severe and frequent sickness. This came after Cotzias's discovery of the value of levo-dopa in Parkinson's disease (Cotzias et al. 1967). Many of our patients, who obtained their initial supplies from London emporia and at considerable expense, were unwilling to abandon the experiment, but sat by their sinks for the initial months of treatment, yawning, pale and retching, immobilised by sickness, despite improvement in the Parkinsonism. Guggenheim, who first isolated Dopa from broad beans grown in the Hoffman La Roche factory fields, tested the compound for pharmacological activity on a variety of preparations, but found it completely inert. Finally, he took 2.5 g pure Dopa himself, which produced copious vomiting, although scientific to the last, he recorded the detailed results in his notebook (Klingler 1982). Despite the advent of decarboxylase inhibitors, which reduce the incidence of levo-dopa-induced vomiting at least fourfold (Marsden and Parkes 1977), levo-dopa-induced sickness remains a major problem in the control of Parkinsonism, and of a severity which would only be tolerated otherwise in the drug treatment of malignant disease. Vomiting produced by levo-dopa and dopamine agonists is usually attributed to a central, and not a gastric, effect (Loew et al. 1978; Agid et al. 1979; Bateman et al. 1980; Quinn et al. 1981; Gonce et al. 1982; Shindler et al. 1984).

Headache and Vomiting

The common association of gastro-intestinal symptoms with headache has been recognised since at least the time of Hippocrates; and in most instances, the complaint of sickness to a neurologist comes from patients with migraine. In classic migraine, nausea and vomiting are the most troublesome abdominal symptoms, although the underlying mechanisms are not well understood, and in some other forms of migraine (e.g. migrainous neuralgia), gastro-intestinal disturbances are not a feature. The headaches of temporal arteritis, severe tension, cervical spondylosis, do not cause sickness in the average subject. Nausea in migraine is often associated with pallor and tachycardia (sometimes bradycardia), which are likely to result from excessive stimulation of the autonomic nervous system; and perhaps result from disturbed cerebral blood flow in acute migraine, but gastro-intestinal factors are likely also to be involved (Volans et al. 1975). Gastric stasis has been shown to occur in acute migraine. This, rather than the pain of migraine, is probably the major cause of nausea, and possibly of vomiting.

Raised Intracranial Pressure

Projectile vomiting, sometimes not preceded by nausea, and often most severe in the early morning, is the commonest presentation of sickness with raised intracranial pressure, usually considered to result from changes in blood flow, permeability and function of the brain-stem vomiting mechanisms. Here, the absence of nausea may be associated with the absence of reflex factors or gastric stasis. Relief of cerebral oedema associated with malignant tumours by corticosteroids or by operation usually relieves sickness within 24–48 h.

Giddiness and Sickness

Acute labyrinthine vertigo (but not minor degrees of persistent unsteadiness) is a common cause of both nausea and vomiting. One of the most giddy patients I have encountered, a hard-working general practitioner, was also one of the most sick. At the time of an attack of vestibular neuronitis, he was only able to crawl, snake-like, along the floor, more serious motion provoking collapse, giddiness, and rapid stomach emptying. Anticholinergic drugs (hyoscine), but not dopamine antagonists (domperidone, metoclopramide), relieved his symptoms – stressing the importance of correct diagnosis and consideration of cause in the treatment of any form of sickness.

Reflex Vomiting

Reflex mechanisms are probably far more common causes of vomiting than disease of the central nervous system (CNS). Cortical as well as gastro-intestinal reflex systems are involved, although, of course, stomach irritation with food intolerance or toxicity is, in most people, more frequent than vomiting due to cortical stimulation, intense motion, high arousal or sustained anxiety. All of these do cause sickness, but with the exception of anorexia nervosa, continued sickness is not a common presenting symptom of any of the major psychiatric disorders. In the sickness of pregnancy, abdominal pressure afferents, not the fear of childbirth, may possibly determine the occurrence of sickness.

The Brain-Stem Vomiting Centre

These clinical examples suggest the presence of at least five different pathways to the brain-stem vomiting centre which coordinates the motor response with yawning, inspiration, closure of the glottis, and forced retrograde gastric emptying, that make up the act of vomiting.

These afferent pathways include:

1. Gastro-medullary afferents which are carried in the vagus
2. Afferents from other abdominal viscera and peritoneum
3. A cortico-bulbar system, triggered mainly by apprehension and fear
4. Vestibular afferents
5. Chemosensor systems connected to the vomiting centre, usually considered to lie outside the blood-brain barrier

It is readily apparent that the stimulation of vomiting by apprehension is not primarily dependent on arousal. Following Moruzzi's (1972) view of arousal as a continuum from the lowest state, deep coma, to hyperalertness, rage and mania, any sudden shift in the level of arousal is not accompanied by either sickness or vomiting. The nausea which is occasionally produced by central stimulant drugs (e.g. amphetamine, methylphenidate, mazindol) is independent of the alerting effect of these agents, and perhaps conditioned by either gastric reflex or dopamine-chemoreceptor stimulation. Clearly, some other explanation for "cortical vomiting" is involved. Sometimes this response is conditioned by gustatory, olfactory or visual stimuli (e.g. sickness at the sight of specific foods); less often it is due to non-specific stimuli. Such conditioned mechanisms may respond to aversion therapy, but are rarely of importance in neurology.

Nausea and vomiting, if sustained, almost always produce anorexia. Vomiting induced by the intense fear of gaining weight and becoming obese is not, however, uncommon, particularly with periodic bulimia that occurs at some stages of illness in approximately half of all anorectics.

Vomiting Centre in Man

There is evidence for the presence of a low medullary vomiting centre in man located in the floor of the medulla in the neighbourhood of the 9th–12th cranial nerve nuclei, although there is more support for the idea of an anatomically separate, but functionally connected, chemoreceptor trigger zone (CTZ). Persistent severe vomiting is frequent in tumours of the posterior fossa, and in patients with small, low, focal brain-stem gliomas. Early in this century, surgeons found that low medullary handling was not without danger. Medullary puncture at open operation or compression could produce intense angor animi, nausea and respiratory arrest. Electrical stimulation of the human brain has usually been confined to cortical areas, and vomiting is not amongst the usual sequelae: it is unusual for small cortical lesions by themselves to produce sickness. However, small discrete brain-stem lesions, infarctions or tumours, as well as causing raised intracranial pressure, are not infrequently associated with hiccup, yawning and persistent vomiting. Some patients may vomit unexpectedly and forcibly, without any preceding nausea, but the majority have both nausea and vomiting. In contrast, nausea and vomiting are not features of progressive brain-stem degenerative disorders (e.g. progressive supranuclear palsy, multisystem atrophy). In progressive supranuclear palsy, as well as motor neurone disease, despite very considerable loss of brain-stem neurons, levo-dopa sometimes provokes considerable nausea and vomiting.

Good evidence for a central mechanism of emesis in man is given by the frequent provocation of vomiting by any sudden rise in intracranial pressure (e.g. intracerebral or subarachnoid haemorrhage). In contrast, sudden hemisphere thrombosis or embolism, without increase in intracranial pressure, very rarely causes sickness. The evidence for an anatomically separate CTZ situated outside the blood-brain barrier but inside the brain stem in man is not conclusive. For example, intracarotid cytotoxic drugs, used in the treatment of cerebral glioma, and which may not immediately penetrate posterior vascular territory, not infrequently are followed by nausea. However, here the situation is complicated by cortical factors, other drugs, and often simultaneous deep X-ray therapy (DXT). It is still possible that the major CTZ in man lies in the stomach, not in the area postrema (AP) of the fourth ventricle.

Pharmacologically, emetic as well as anti-emetic drugs have very comparable effects between species: ergot alkaloids, for example, cause vomiting in dogs and monkeys, as well as in man, and similar mechanisms are probably applicable in different species. However, it is not always possible to dissect out separate gastric and central effects of emetic and anti-emetic drugs in man with the data presently available.

Neurological Drug Therapy and Vomiting

Emetic Drugs and the Motor System

It is perhaps surprising that most, if not all, the dopamine-like drugs used in the management of Parkinsonism also cause frequent vomiting, whilst dopamine antagonists, which cause many and serious motor side effects, are mainly anti-emetic in their effects. Drug receptors in the central motor, as well as emetic systems, are stimulated by dopamine, but with little or no selectivity of action; all effective dopamine-like drugs in the treatment of Parkinsonism will produce sickness, although, at least with the ergot alkaloids, the emetic potency in animals and the effects on the motor system (locomotor hyperactivity, turning in the Ungerstadt model in the rat) are unrelated. Loew et al. (1978) explained the difference on the basis of different levels of accessibility, different species and routes of administration of ergot drugs acting on the emetic CTZ. Tolerance to the emetic effect of dopamine agonists usually, but not always, develops with continued treatment.

Gastric Effects of Dopamine-Like Drugs

The possibility that gastric stimulation may be responsible for nausea and vomiting, due to anti-Parkinsonian drugs, has been neglected until the recent demonstration of stomach dopamine receptors and the finding that levo-dopa, itself, causes marked gastric stasis. This, in turn, may result in reflex nausea. In a few individual subjects with a vagotomy for pyloric stenosis who were given levo-dopa for the treatment of Parkinsonism, this drug did not produce any sickness (N. Langdon 1984, personal communication), but the relative importance of gastric and central mechanisms remains undecided (decarboxylase inhibitors will prevent formation of dopamine in the stomach wall as well as in a postulated CTZ). As with the motor-emetic effects of ergot alkaloids, the different motor and anti-emetic effects of dopamine antagonists show little correlation, although direct comparisons for anti-emetic potency and liability to cause, for example, Parkinsonism, are lacking with most of the phenothiazines, butyrophenones and substituted benzamides. Although drug-induced Parkinsonism is the likely consequence of nigrostriatal dopamine receptor blockade, the exact mechanism of the other motor side effects with anti-emetic phenothiazine drugs (e.g. akathisia, acute dystonia and chronic or tardive dyskinesia) remains obscure. Anticholinergic anti-emetic drugs (e.g. hyoscine used in labyrinthine vomiting) do not produce these effects in conventional dosage, although hyoscine toxicity sometimes does produce chorea.

Domperidone and Sickness

Domperidone is an important exception to the general rule that dopamine antagonist anti-emetic drugs prevent sickness, but also cause motor disorders. Although domperidone has been associated with isolated instances of acute dystonia (Sol et al. 1980; Debontrïdder 1980a, b; Gonce et al. 1982; Van Daele et al. 1984), 13 other motor side effects have not been reported (unlike metoclopramide), probably owing to the absence of brain penetration by domperidone in conventional dosages. Thus, domperidone, like decarboxylase inhibitors, can be used as a successful anti-emetic with levo-dopa or other dopamine agonist drugs, including the ergot alkaloids. Domperidone does not have the dopa-sparing potential of decarboxylase inhibitors, but like these drugs may protect against dopamine-induced cardiac arrhythmias, as well as against sickness (N. P. Quinn 1984, personal communication). Also, by speeding gastric emptying, domperidone–levo-dopa combinations abolish the multiple peaks of dopa absorption seen with levo-dopa alone, and may slightly increase initial levo-dopa bio-availability following oral levo-dopa dosage (Shindler et al. 1984). This effect is, however, very small compared with that of decarboxylase inhibitors. Domperidone protects against levo-dopa-, lisuride-, apomorphine-, pergolide- and bromocriptine-induced vomiting, but does not apparently prevent hiccup or yawning induced by levo-dopa (Quinn et al. 1981).

References

Agid Y, Pollak P, Bonnet AM, Signoret JL, Lhermitte F (1979) Bromocriptine associated with a peripheral dopamine blocking agent in treatment of Parkinson's disease. Lancet 1:570–572

Bateman DN, Kahn C, Davies DS (1980) The pharmacokinetics of metoclopramide in man with observations in the dog. Br J Clin Pharmacol 9:371–377

Cotzias GC, Van Woert MH, Schiffer LM (1967) Aromatic amino acids and modification of parkinsonism. N Engl J Med 276:374–379

Debontrïdder O (1980a) Extra-pyramidal reactions due to domperidone. Lancet 2:802

Debontrïdder O (1980b) Dystonic reactions after domperidone. Lancet 2:1259

Gonce M, Bury J, Burton L, Delwaide PJ (1982) Syndrome neurodysleptique induit par le domperidone. Nouv Presse Med 11:2298

Klingler M (1982) A glimpse into Roche research: antiparkinsonian treatment. Int J Clin Pharmacol Ther Toxicol 20:190–193

Loew DM, Van Deusen EB, Meier-Ruge W (1978) Effects on the central nervous system. In: Berde B, Schild HO (eds) Ergot alkaloids and related compounds. Springer, Berlin Heidelberg New York, pp 421–531

Marsden CD, Parkes JD (1977) Success and problems of long-term levodopa therapy in Parkinson's disease. Lancet 1:345–349

Matthews B (1963) Practical neurology. Blackwell, Oxford

Moruzzi G (1972) The sleep-waking cycle. Ergeb Physiol 64:1–165

Quinn N, Illas A, Lhermitte F, Agid Y (1981) Bromocriptine and domperidone in the treatment of Parkinson's disease. Neurology (NY) 31:662–667

Shindler J, Finnerty G, Jenner P, Lake-Bakaar G, Parkes JD (1984) Domperidone, metoclopramide and levodopa in Parkinson's disease. Proceedings of First Symposium of the European Neuroendocrine Association, 4–7 March, Basle, p 42

Sol P, Pelet B, Guignard JP (1980) Extrapyramidal reactions due to domperidone. Lancet 2:802

Van Daele CM, Dobosz-Cyklis R, Van de Cassege W, Verbeek P, Wijndaele L (1984) Refusal of further cancer chemotherapy due to antiemetic drug. Lancet 1:57

Volans GN (1975) The effect of metoclopramide on the absorption of effervescent aspirin in migraine. Br J Clin Pharmacol 2:57–63

Rational Approaches to Therapy for Nausea and Vomiting: Two Common Solutions

M. D. Rawlins [1], W. B. Taylor [1], and D. N. Bateman [1]

Nausea and vomiting are associated with many diseases and pathological processes. They are probably most often encountered, however, in association with gastro-intestinal tract infections, influenza and other self-limiting febrile disorders. The anti-emetic drugs most commonly prescribed for symptomatic relief in these circumstances are metoclopramide and prochlorperazine, and their rational use would be most reasonably based on knowledge of their pharmacological and pharmacokinetic properties, and on clear evidence of their safety and efficacy for these indications.

The Pharmacology of Metoclopramide and Prochlorperazine

Bot metoclopramide and prochlorperazine are dopamine antagonists but, whilst metoclopramide shows selectivity for D_2 receptors prochlorperazine is active at both D_1 and D_2 receptors (Pinder et al. 1976; Kebabian and Calne 1979). Metoclopramide also has cholinomimetic actions in the gastro-intestinal tract (Hay 1977) by influencing the release of acetylcholine (Hay and Man 1979) possibly via antagonism of presynaptic cholinergic receptors (Fosbraey et al. 1980). In vitro metoclopramide has also been shown to possess anti-serotoninergic properties (Kilbinger et al. 1982), but these activities can only be observed in vitro at concentrations considerably in excess of those usually observed in vivo. Prochlorperazine in common with other phenothiazines has antihistaminergic (H_1), anticholinergic and anti-adrenergic (alpha$_1$) properties (Baldessarini 1980), but it is uncertain whether these contribute to its therapeutic effects.

In man, therapeutic doses of both drugs release prolactin as a result of their dopamine-antagonist actions. In addition, metoclopramide increases human gastro-intestinal motility, probably by its action on cholinergic transmission in the gut. Consequently, metoclopramide increases gastric emptying rates and reduces small intestinal transit times (Harrington et al. 1983).

Pharmacokinetics

Little is known of the pharmacokinetics of prochlorperazine. Sankey et al. (1982) have reported plasma concentrations of less than 1 μg/l after the oral administra-

1 Wolfson Unit of Clinical Pharmacology, Department of Pharmacological Sciences, University of Newcastle-upon-Tyne, NE1 7RU, UK

Nausea and Vomiting: Mechanisms and Treatment
Ed. by C. J. Davis, G. V. Lake-Bakaar and D. G. Grahame-Smith

tion of 12.5 mg prochlorperazine, but there is no published information available concerning its bio-availability, distribution, half-life or clearance.

The pharmacokinetics of metoclopramide have recently been reviewed (Bateman 1983). Although it is well-absorbed from the gastro-intestinal tract, its bio-availability is incomplete and variable after oral administration, ranging from 32% to 97% (Bateman et al. 1980). Its elimination half-life is 3–5 h (Bateman et al. 1980), with an overall clearance of $0.70 \pm$ SD 0.21 l/kg/h. Approximately 20% of an intravenous dose of metoclopramide is excreted unchanged in the urine; a sulphate conjugate accounts for a further 32%, and small quantities of glucuronide conjugate as well as a product of side-chain cleavage (4-amino-5-chloro-2-methoxy-benzamide-acetic acid) have also been identified as metabolites in man.

In children receiving high doses of metoclopramide ($0.35 \pm$ SE 0.025 mg/kg) as prophylaxis against cytotoxic-induced vomiting, Bateman et al. (1983) observed that the clearance ($0.56 \pm$ SD 0.10 l/kg/h) was similar to that of adults, and that plasma concentrations in three children who developed dystonic conditions were similar to those of six children who did not. In renal failure, however, clearance is markedly reduced ($0.26 \pm$ SD 0.08 l/kg/h) and the half-life is prolonged to 14 h (Bateman et al. 1981); this may explain the increased incidence of adverse reactions to the drug reported in patients with impaired renal function (Caralps 1979).

Whilst evidence for a direct correlation between plasma concentrations of metoclopramide and its anti-emetic efficacy is sparse, there are indications that such a correlation may exist for its adverse central nervous system (CNS) effects (Bateman 1983). This important area, however, requires further investigation.

Efficacy

Metoclopramide and prochloroperazine are used in a wide variety of disorders. The indications for metoclopramide include dyspepsia, flatulence, heartburn, peptic ulceration, gastritis, cholelithiasis, migraine, diagnostic procedures and nausea and vomiting associated with a range of conditions (Data Sheet Compendium 1983). Most of these indications can be supported by evidence of efficacy from controlled clinical trials (Schulze-Delrieu 1979; Harrington et al. 1983), but we are unaware of good published information which confirms its therapeutic value in what is probably the commonest indication – the nausea and vomiting accompanying acute gastro-intestinal tract infections and similar self-limiting conditions.

Prochlorperazine is advocated (Data Sheet Compendium 1983) for the treatment of Menieres disease, labyrinthitis, migraine, schizophrenia and nausea and vomiting "from whatever cause". Again, although there is evidence for the anti-emetic efficacy of prochlorperazine in many cases of nausea and vomiting, we are unaware of convincing clinical data which demonstrates its efficacy in gastro-intestinal infections.

Safety

Prochlorperazine and metoclopramide cause predominantly CNS adverse effects (Table 1) and Robinson (1973) reported that, for metoclopramide, drowsiness occurred in 5% to 10% of patients receiving the drug. With both compounds, probably the most dramatic adverse CNS effects are the extra-pyramidal reactions

Table 1. Adverse reactions to metoclopramide and prochlorperazine reported to the Committee on the Safety of Medicines (1964–1984)

Reaction	Metoclopramide	Prochlorperazine
Total cases	935 (4)	306 (7)
Skin	52 (1)	38 (0)
Central nervous system	720 (0)	168 (1)
Autonomic nervous system	21 (0)	8 (0)
Reproduction system	41 (0)	0 (0)
Gastro-intestinal	19 (0)	18 (0)
Liver	1 (0)	18 (4)
Blood	5 (1)	11 (1)
Others	76 (2)	45 (1)

Figures in parenthesis indicate the number of reported deaths

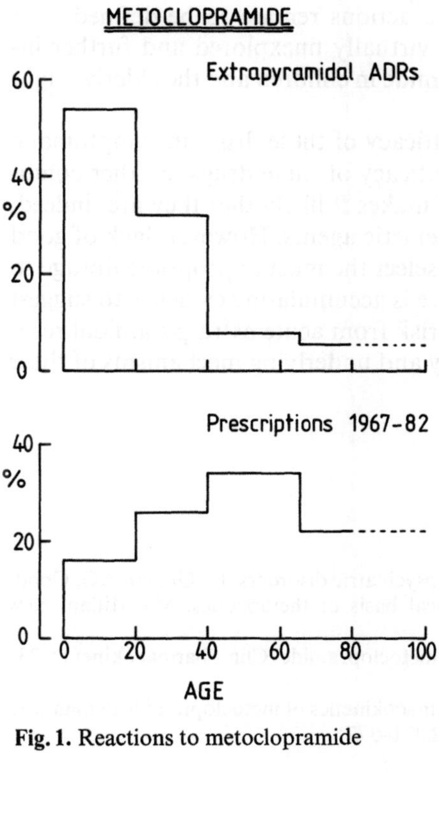

Fig. 1. Reactions to metoclopramide

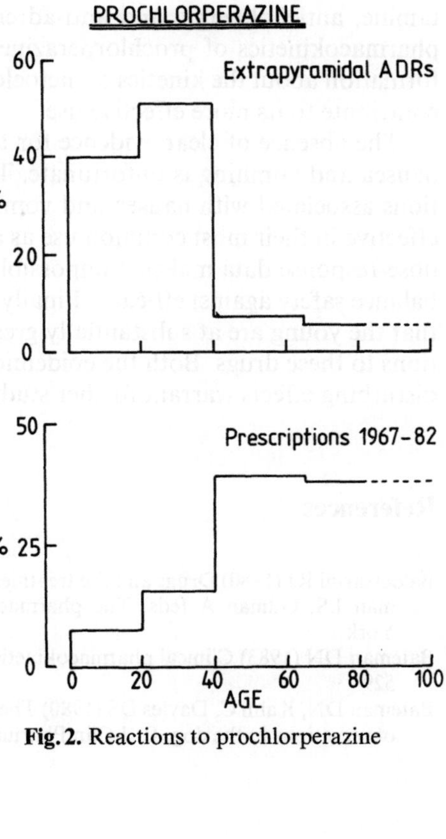

Fig. 2. Reactions to prochlorperazine

presenting as acute dystonias or dyskinesias. These constitute the majority of adverse CNS reactions to the two drugs reported to the United Kingdom Committee on the Safety of Medicines (549 and 131 metoclopramide and prochlorperazine, respectively). Casteels-van Daele et al. (1970) suggested that children might be more susceptible to the acute extra-pyramidal effects of metoclopramide, and analysis of the age distribution of adverse reaction reports (ADRs) to the Committee on the Safety of Medicines supports this and the excess of reports in the young cannot be attributed to increased prescribing in this age group (Fig. 1). Similarly (Fig. 2), there is an excess number of extra-pyramidal reactions to prochlorperazine in young children which cannot be accounted for by prescribing. The possibility of reporting bias cannot, however, be entirely discounted and there is an urgent need for prospective studies to establish the incidence of the reaction in relation to dosage, age and sex.

Discussion

It is clear that the widespread use of metoclopramide and prochlorperazine for the treatment of symptomatic nausea and vomiting in ineffective gastro-intestinal disorders does not rest on an entirely rational basis. Although the human pharmacology of metoclopramide appears to have been reasonably well-defined, that of prochlorperazine is unclear; in particular, the relevance of its possible antihistamine, anticholinergic and anti-adrenergic actions remain to be defined. The pharmacokinetics of prochlorperazine are virtually unexplored and further information about the kinetics of metoclopramide in children and the elderly might contribute to its more effective use.

The absence of clear evidence for the efficacy of these drugs in symptomatic nausea and vomiting is unfortunate. The efficacy of these drugs in other conditions associated with nausea and vomiting makes it likely that they are, indeed, effective in their most common use as anti-emetic agents. However, lack of good dose-response data makes it impossible to select the most appropriate dosage to balance safety against efficacy. Finally, there is accumulating evidence to suggest that the young are at substantially greater risk from acute extra-pyramidal reactions to these drugs. Both the epidemiology and underlying mechanisms of these disturbing effects warrant further study.

References

Baldessarini RJ (1980) Drugs and the treatment of psychiatric disorders. In: Gilman AG, Goodman LS, Gilman A (eds) The pharmacological basis of therapeutics. MacMillan, New York

Bateman DN (1983) Clinical pharmacokinetics of metoclopramide. Clin Pharmacokinet 8:523–529

Bateman DN, Kahn C, Davies DS (1980) The pharmacokinetics of metoclopramide in man with observations in the dog. Br J Clin Pharmacol 9:371–377

Bateman DN, Gokal R, Dodd TPR, Blain PG (1981) The pharmacokinetics of single doses of metoclopramide in renal failure. Eur J Clin Pharmacol 19:437–441
Bateman DN, Craft AW, Nicholson E, Pearson ADJ (1983) Dystonic reactions and the pharmacokinetics of metoclopramide in children. Br J Clin Pharmacol 15:560–563
Caralps A (1979) Metoclopramide and renal failure. Lancet 1:554
Casteels-van Daele M, Jaeken J, van der Schueren P, Zimmermann A, van den Bon P (1970) Dystonic reactions in children caused by metoclopramide. Arch Dis Child 45:130–133
Data Sheet Compendium (1983) Datapharm Publications, London
Fosbraey P, Hird MF, Johnson ES (1980) The effects of some dopamine antagonists on cholinergic mechanisms in the guinea pig ileum. J Auton Pharmacol 1:17–25
Harrington RA, Hamilton CW, Brogden RN, Linkewich JA, Romankiewicz JA, Heel RC (1983) Metoclopramide – an update review of its pharmacological properties and clinical use. Drugs 25:451–494
Hay AM (1977) Pharmacological analysis of the effects of metoclopramide on the guinea pig isolated stomach. Gastroenterology 72:864–869
Hay AM, Man WK (1979) Effect of metoclopramide on guinea pig stomach: critical dependence on intrinsic stores of acetylcholine. Gastroenterology 76:492–496
Kebabian JM, Calne DB (1979) Multiple receptors for dopamine. Nature 277:93–96
Kilbinger H, Kruel R, Pfeuffer-Frederick I, Wessler I (1982) The effects of metoclopramide on acetylcholine release and on smooth muscle response in the isolated guinea-pig ileum. Naunyn Schmiedebergs Arch Pharmacol 319:231–238
Pinder RM, Bogdea RM, Sawyer PR, Speight TM, Avery GS (1976) Metoclopramide: a review of its pharmacological properties and clinical use. Drugs 12:81–131
Robinson OPW (1973) Metoclopramide – side effects and safety. Postgrad Med J 49 [Suppl 4]:77–80
Sankey MG, Holt JE, Kaye CM (1982) A simple and sensitive HPLC method for the assay of prochlorperazine in plasma. Br J Clin Pharmacol 13:578–580
Schulze-Delrieu K (1979) Metoclopramide. Gastroenterology 77:768–779
Schulze-Delrieu K (1981) Metoclopramide. N Engl J Med 305 (1):28–33

Clinical Discussions: An Overview

G. V. Lake-Bakaar [1]

The discussion sessions which followed the clinical papers formed an integral part of the Symposium. They provided an opportunity for discussion of the papers presented by the speakers, and for participants to present relevant items of original research. It is not practical to provide a full transcript of these sessions here. This chapter aims merely to provide the "flavour" of these discussions which often were as interesting, pertinent and revealing as the papers themselves.

No attempt will be made to attribute specific remarks to members of the audience. The main aim will be to provide a personal view of the several very fascinating and often lively debates.

One question which recurred throughout the meeting was simply: Why vomit? What possible value does vomiting confer upon the survival of the species? This question was addressed during the discussion on motion sickness. Motion sickness or its equivalent can be induced in a wide variety of animals including birds and fish. The idea of fish swimming through their own vomitus evoked a great deal of mirth! Sensory information required for motion is derived from multiple neural sources and needs to be highly integrated. Ingestion of neuro-toxic substances might, for example, readily upset the balance. Therefore, if the emetic response were linked in some way with this system, it would provide a sensitive sensor mechanism and thus confer survival advantage.

Is motion sickness linked with any of the other clinical conditions which provoke nausea and vomiting? A possible association with post-operative vomiting was suggested by one speaker from the floor. Motion-sickness susceptibility is of predictive value for susceptibility to post-operative nausea and vomiting. The two conditions may be closely related. Motion is involved in the "bumpy" ride to the theatre; at the end of the anaesthetic the patient is turned to her side; she is then bounced back to the ward on a trolley and hurled on to her bed; finally, if she has not vomited she gets up suddenly to telephone her relatives – and then vomits!

A major stimulus to peri-operative nausea and vomiting must, of course, also be due to chemical interference which is akin to the ingestion of neurotoxic substances. A study of the effect of minimising or eradicating motion during anaesthesia on the incidence of nausea and vomiting appears to be indicated.

The general question of methodology was also discussed. The point was made that competent investigators using identical materials obtained contradictory findings. A plea for greater standardisation of methodology was made. In the area of post-operative nausea and vomiting, investigators tended to study vomit-

1 Medical Department Janssen Pharmaceutical Limited, Grove, Wantage, Oxfordshire OX12 0DQ, UK

Nausea and Vomiting: Mechanisms and Treatment
Ed. by C. J. Davis, G. V. Lake-Bakaar and D. G. Grahame-Smith

ing alone. Although vomiting has the advantage of being a crisp and objective end-point, nausea is often more distressing. In view of post-operative amnesia, patients often needed to be studied for 24 h; this is wasteful of research time. The use of a microphone to pick up sounds from the glottis during retching was suggested, so that patients do not need to be observed continuously.

One speaker expressed concern that the symposium had not addressed the issue of "which neuro-transmitter was involved in which emetic pathway". It was suggested that clinical trials might be organised to try to solve this. The speaker hypothesised that a pure selective H_1 histamine blocker might be better than dopamine blockade in motion sickness, but be less effective in radiation sickness. This point was hotly debated; the many limitations of clinical trials in this setting were aired. It was pointed out that systematic trials in radiation-induced emesis in dogs had been performed with different classes of drugs and with combinations of drugs. This did not appear to have solved the problem. Drug combinations increased the threshold for vomiting; once this threshold was exceeded however, the severity of the ensuing emesis was little altered.

The debate following the paper by A. L. Harris centred around his hypothesis that cytotoxic agents induced nausea and vomiting by interference with DNA and protein synthesis. This resulted in decreased production of the enzymes responsible for the breakdown of the neurotransmitters which produced emesis. The hypothesis was clearly novel, interesting and testable; it answered some of the questions surrounding cytotoxic-induced nausea and vomiting including the sometimes very prolonged delay between drug administration and emesis. There remained many unsolved questions; however, the debate certainly provoked a stimulus to finding answers.

R. W. Young suggested in his talk that radiation-induced nausea and vomiting was at least in part associated with brain oedema and inflammation. Thus, steroids, glycerol and anti-inflammatory drugs conferred some benefit. Nausea and vomiting also followed gut irradiation and the possibility of emesis-mediating substances was raised. The possible role of the recently described intestinally derived emetic factor was discussed. Histamine was unlikely to be a major candidate since (a) antihistamines are not totally effective against nausea and vomiting and (b) gastric acid output decreased rather than increased following gut irradiation. The interesting possibility of using cross-circulation experiments to test for possible mediators also arose.

J. R. R. Stott was asked about the differences and similarities between terrestrial motion sickness and space motion sickness. First, both exhibit the same adaptive type of response; after a few days on stormy seas, people adapt – the same situation occurs in space. There is also the same latency and relationship to head movements in space. It was of interest that space sickness only became a problem when motion within the space capsule became much more free. One area of difference, however, was that whereas it was possible to predict susceptibility to terrestrial motion sickness, predictive tests for space sickness are not available. This appeared to be a significant problem, since 40% to 50% of subjects developed symptoms of space motion sickness.

The question of scoring methods for assessing nausea and vomiting post-operatively was raised following R. F. Cookson's paper. Should nausea, retching and vomiting be scored separately? Are all three symptoms on the same continuum? Should nausea during the first post-operative hour be scored separately from that after hours 1 to 6?

Another important point was raised concerning the distress caused by nausea. Nausea is subjective and difficult to measure; whereas, vomiting is objective and reasonably easy to quantify. Reducing the number of episodes of vomiting from ten to two may not be such a major advance if the patient remains nauseated. Moreover, it is often the case that nauseated patients felt much better after vomiting.

The importance of treatment with anti-emetic agents was also discussed, especially with regard to cancer chemotherapy. The comment was sometimes made "I stopped taking the medicine because I prefer the original disease to the side effects". Given that highest-tolerated doses of cancer chemotherapeutic agents influence survival time, nausea and vomiting is fast becoming a significant challenge to survival for cancer patients. The increasing importance of nausea and vomiting is exemplified by the recent increase in the number of anti-emetics trials in humans.

Several interesting points were raised in the debate following the paper by D. V. I. Fairweather. For example, was hyperemesis gravidarum a separate entity or was it merely an exaggerated form of the physiological morning sickness? Is it a neurotic manifestation against an unwanted pregnancy, in which case hyperemesis might be expected to be related to abortion policies? Unfortunately, data presented by D. V. I. Fairweather predated the Abortion Acts in the UK. They suggested major geographical differences in the incidence of hyperemesis; the condition was commoner in Aberdeen and northeast England, but was relatively uncommon in the south of England. Nowadays, the condition has practically disappeared. This might account for the paucity of recent literature in this area.

The mechanism of action of decarboxylase inhibitors and the peripheral dopamine antagonist domperidone in preventing L-dopa-induced nausea and vomiting was the subject of debate between J. D. Parkes and the Chairman, D. G. Grahame-Smith. Certain points were highlighted. Do the peripheral dopamine (PD) inhibitors diminish the vomiting due to L-dopa because the "load" of L-dopa is diminished (thereby causing less gastro-intestinal "irritation"), or because there is less dopamine in the peripheral blood or within the AP to produce chemical emesis? It is plainly important to know whether the peripheral decarboxylase inhibitors block aromatic amino-acid decarboxylase in the AP. It could also be that peripheral decarboxylase inhibition blocks the formation of dopamine in the gastrointestinal wall which could conceivably be emetic. Domperidone is effective against L-dopa-induced vomiting whilst not affecting its antiparkinsonian effect. Presumably it is acting as a dopamine antagonist at the AP or peripherally, but as it does not pass the blood barrier, not in the basal ganglia.

The effect of suggestion in clinical experience was discussed following M. D. Rawlins' paper. A major difference exists between: "This will take care of it" vs "Why don't you give this a try and see if it helps?". This might go some way towards explaining the divergent results by various authors in clinical trials of intravenous anti-emetic agents.

There was little doubt that the standard of the papers presented, both during the basic science day and the clinical medicine day, was high. Some areas, however, especially in clinical medicine, have been sadly neglected over the years; this includes pregnancy and neurology, where the incidence of serious symptoms is arguably low. A major stimulus to clinical research and the search for better antiemetic drugs has come from the area of oncology, where emesis can actually limit therapy. The hope was expressed that this interest would continue.

Participants

Dr. P. L. R. Andrews

Department of Physiology
St. George's Hospital Medical School
University of London, Cranmer Terrace
Tooting, London SW17 ORE, UK

Professor A. Angel

Department of Physiology
The University of Sheffield
Sheffield, S10 2TN, UK

Dr. J. K. Aronson

Reader in Clinical Pharmacology
MRC Clinical Pharmacology Unit and
University Department
of Clinical Pharmacology
Radcliffe Infirmary, Woodstock Road
Oxford, OX2 6HE, UK

Mr. J. H. Barnes M. Sc.

1 Sims Gardens, Main Street,
Grove
Wantage, Oxfordshire, OX12 7LF, UK

Dr. A. Barrett

Consultant in Radiotherapy and Oncology
The Royal Marsden Hospital
Downs Road, Sutton
Surrey SM2 5PT, UK

Dr. J. J. Brand

31 The Avenue, Alverstoke
Gosport, Hampshire, UK

Professor K. R. Brizzee

Department of Neurobiology
Tulane University, Covington
Louisiana 70433, USA

Dr. D. O. Carpenter

Director, Centre for Laboratories
and Research, State of New York Office of
Public Health
Corning Tower, Albany, New York 12201,
USA

Dr. R. F. Cookson

Clinical Development Manager
Medical Department
Janssen Pharmaceutical Limited
Grove, Wantage
Oxon OX12 0DQ, UK

Dr. D. J. Costello

Assistant Director
Clinical Information
Hoechst-Rousell Pharmaceuticals Inc
Route 202-206 North, Somerville
New Jersey 08876, USA

Professor G. H. Crampton

Professor of Psychology
Department of Psychology
Wright State University
Dayton, Ohio 45435, USA

Dr. J. Curran

Department of Anaesthesia
Floor C, West Block, University Hospital
Nottingham NG7 2UH, UK

Dr. N. Daunton

Biomedical Research Division N239-7
NASA Ames Research Center
Moffett Field, California 94035, USA

Surgeon Lieutenant Commander
C. J. Davis Royal Navy

MRC Clinical Pharmacology Unit and
University Department
of Clinical Pharmacology
Radcliffe Infirmary
Woodstock Road, Oxford OX2 6HE, UK

Dr. R. Drawbaugh

Chief, Biotechnology
U.S. Airforce Office of Scientific Research
European Office of Aerospace
Research & Development
223/231 Old Marylebone Road
London NW1 5TH, UK

Dr. A. Dubois, Assistant Director

Digestive Diseases Division
Department of Medicine
Uniformed Services University
of the Health Sciences
Bethesda Maryland 20814-4799, USA

Professor D. V. I. Fairweather

Professor of Obstetrics and Gynaecology
Department of Obstetrics and Gynaecology
Faculty of Clinical Sciences
University College Hospital
School of Medicine
University of London
88–96 Chenies Mews, London WC1E 6HX, UK

Dr. R. H. Freeman

Consultant Physician
Macmillan Service
King Edward VII Hospital
Midhurst, Sussex, GU29 OBL, UK

Surgeon Commander
F. St. C. Golden OBE Royal Navy
Director of Research
Institute of Naval Medicine
Alverstoke, Gosport, Hampshire PO12 2DL, UK

Professor D. G. Grahame-Smith

MRC Clinical Pharmacology Unit and
University Department
of Clinical Pharmacology
Radcliffe Infirmary
Woodstock Road, Oxford OX2 6HE, UK

Dr. A. R. Green

Assistant Director
MRC Clinical Pharmacology Unit and
University Department
of Clinical Pharmacology
Radcliffe Infirmary
Oxford OX2 6HE, UK

Dr. S. J. Gunning

Postgraduate School for the Study
of Pharmacology
University of Bradford
Bradford, West Yorkshire BD7 1DP, UK

Professor D. G. Gwyn

Professor and Head of Department
Department of Anatomy
Dalhousie University
Halifax, Nova Scotia B3H 4HT, Canada

Dr. J. A. Gylys

Associate Director
Research & Development Division
Bristol-Myers Pharmaceutical Co
PO Box 4755, Syracuse
New York 13221-4755, USA

Dr. R. K. Harding

Nuclear Effects Section, Protective
Sciences Division, Defence Research
Establishment Ottawa, Department
of National Defence, Shirley Bay, Ottawa
Ontario, KIA OZA, Canada

Professor A. L. Harris

University Department of Radiotherapy
Newcastle University Medical School
Newcastle General Hospital Westgate Road
Newcastle-upon-Tyne NE4 6BE, UK

Dr. C. J. Hawkey

Senior Lecturer in Therapeutics
Department of Therapeutics
Floor C, South Block, University Hospital
Nottingham NG7 2UH, UK

Dr. M. Herbert

Senior Lecturer in Behavioural Sciences
Department of Psychiatry
University of Nottingham Medical School
Floor A, South Block, University Hospital
and Medical School
Clifton Boulevard
Nottingham NG7 2UH, UK

Dr. T. Hunt

Director, Arthur Rank House and
Hon Lecturer in Medicine
University of Cambridge
Addenbrooke's Hospital
Hills Road, Cambridge CB2 2QQ, UK

Dr. A. Ireland

Gastroenterology Unit
The Radcliffe Infirmary
Oxford OX2 6HE, UK

Dr. G. V. Lake-Bakaar

Director of Medical Affairs
and Research Gastroenterology Physician
Senior Medical Adviser and Research
Physician
Janssen Pharmaceutical Limited
Grove, Wantage, Oxfordshire OX12 0DQ, UK

Dr. I. N. C. Lawes

Department of Anatomy and Cell Biology
The University of Sheffield
Sheffield S10 2TN, UK

Dr. J. M. Letcombe

Unité 109 de Neurobiologie
Centre Paul Broca de l'INSERM
2 ter rue d'Alésia, 75014 Paris, France

Dr. R. A. Lefebvre

J. F. and C. Heymans Institute
of Pharmacology
University of Ghent Medical School
9000 Ghent, Belgium

Dr. J. Lemoine

Department of Physiology
School of Medicine
Faculty of Health Sciences
University of Ottawa
Ontario, Canada

Dr. R. A. Leslie

Department of Anatomy
Faculty of Medicine
Dalhousie University
Halifax, Nova Scotia B3H 4H7, Canada

Air Commodore I. R. Lindsay RAF

Consultant Adviser in Radiobiology
Central Medical Establishment
Royal Air Force Kelvin House
Cleveland Street, London W1P 6AU, UK

Sister J. Lucken

CRC & Wessex Regional Medical
Oncology Unit
Southampton General Hospital
Southampton SO9 4XY, UK

Dr. R. L. Maynard

Superintendent
Medical Division
Chemical Defence Establishment
Porton Down, Salisbury
Wiltshire SP4 0JQ, UK

Dr. F. J. Mills

Senior Medical Adviser
Janssen Pharmaceutical Limited
Grove, Wantage, Oxon OX12 0DQ, UK

Dr. M. C. W. Minchin

MRC Clinical Pharmacology Unit and
University Department
of Clinical Pharmacology
Radcliffe Infirmary
Woodstock Road, Oxford OX2 6HE, UK

Dr. W. D. Miner

Gastrointestinal Diseases Project
Beecham Pharmaceuticals Research Division
Medicinal Research Centre
Coldharbour Road
The Pinnacles, Harlow, Essex CM19 5AD,
UK

Dr. G. R. Morrow

Associate Professor of Oncology
in Psychiatry
Division of Behavioural
and Psychosocial Medicine
Department of Psychiatry
School of Medicine and Dentistry
University of Rochester Medical Center
Stroney Memorial Hospital
Rochester, New York 14642, USA

Dr. C. A. Muir

Department of Biochemistry
University Hospital and Medical School
Clifton Boulevard
Nottingham NG7 2UH, UK

Dr. S. G. Murray

Principal Lecturer
The Royal Military College of Science
Shrivenham, Wiltshire SN6 8LA, UK

Dr. F. Nielson

Janssen AB Leo
Box 941
S-251 09 Helsingborg, Sweden

Dr. D. R. Norfolk

Co-ordinator
Yorkshire Regional Bone Marrow
Transplant Group
St. James's University Hospital
Beckett Street, Leeds LS9 7TF, UK

Dr. J. D. Parkes

Reader in Neurology
Department of Neurology
King's College Hospital Medical School
and Institute of Psychiatry University
of London
Denmark Hill, London SE5, UK

Dr. M. G. A. Palazzo
Department of Anaesthetics
Radcliffe Infirmary
Woodstock Road, Oxford OX2 6HE, UK

Dr. C. Patel
Gastroenterology Unit
Radcliffe Infirmary
Woodstock Road, Oxford OX2 6HE, UK

Surgeon Commander
B. J. W. Pingree Royal Navy
Head of Human Factors Division
Institute of Naval Medicine
Alverstock, Gosport, Hants PO12 2DL, UK

Professor M. D. Rawlins
Professor of Clinical Pharmacology
Wolfson Unit of Clinical Pharmacology
University of Newcastle-upon-Tyne
NE1 7RU, UK

Dr. N. W. Read
Honorary Consultant Physician
Senior Lecturer in Physiology
Clinical Research Unit
H Floor, Royal Hallamshire Hospital
Sheffield S10 2JF, UK

Major General D. M. Roberts QHP
Director of Army Medicine
Ministry of Defence (AMD5)
First Avenue House, High Holborn
London WC1V 6HE, UK

Dr. H. A. Robertson MRC
Neurochemical
Pharmacology Unit
Medical Research Council Centre
Medical School, University of
Cambridge
Hills Road, Cambridge CB2 2QH, UK

Dr. J. E. Schurig
Associate Director
Antitumor Biology Department
Pharmaceutical Research & Development
Division
Bristol-Myers Co, PO Box 4755
Syracuse, New York 13221-4755, USA

Dr. J. Shuurkes
Janssen Pharmaceutica nv
Turnhoutsebaan 30
Beerse B2340, Belgium

Professor J.-C. Schwartz
Unité 109 de Neurobiologie
Centre Paul Broca de l'INSERM
2 ter rue d'Alésia, 75014 Paris, France

Dr. J. Sweetenham
Professor of Medical Oncology
CRC & Wessex
Regional Medical Oncology Unit CF 99
Southampton General Hospital
Southampton SO9 4XY, UK

Dr. J. R. R. Stott
Royal Air Force
Institute of Aviation Medicine
Farnborough, Hampshire, UK

Dr. J. M. Talbot
Senior Medical Consultant
Life Science Research Office
Federation of American Societies for
Experimental Biology
9650 Rockville Pike
Bethesda, Maryland 20814, USA

Dr. M. Treisman
University Department
of Experimental Psychology
South Parks Road
Oxford OX1 3UD, UK

Dr. G. R. Venning
Research & Development Director
Janssen Pharmaceutical Limited
Grove, Wantage, Oxon OX12 0DQ, UK

Miss H. A. Waldron M. Sc.
Research Project Manager
Janssen Pharmaceutical Limited
Grove, Wantage, Oxon OX12 0DQ, UK

Dr. K. Watling
The Neurological Sciences Research Centre
Merck Sharp & Dohme Research Laboratories
Terlings Park, Eastwick Road
Harlow, Essex CM20 2QR, UK

Dr. J. L. Willems
J. F. and C. Heymans Institute
of Pharmacology
University of Ghent Medical School
9000 Ghent, Belgium

Miss K. L. Wood BSc

Department of Physiology
St. Georges Hospital Medical School
University of London
Cranmer Terrace
Tooting, London SW17 ORE, UK

Dr. R. W. Young

Biomedical Effects Directorate
Defense Nuclear Agency
Armed Forces Radiobiology Research Institute
Bethesda, Maryland 20814-5145, USA

Dr. J. Zabara

Department of Physiology & Biophysics
Temple University
3223 North Broad Street
Philadelphia
Pennsylvania 19140, USA

Subject Index

MIX
Papier aus verantwortungsvollen Quellen
Paper from responsible sources
FSC® C105338

If you have any concerns about our products,
you can contact us on
ProductSafety@springernature.com

In case Publisher is established outside the EU,
the EU authorized representative is:
Springer Nature Customer Service Center GmbH
Europaplatz 3, 69115 Heidelberg, Germany

Printed by Libri Plureos GmbH
in Hamburg, Germany